Graphic Design in Television

To Richard my son, who arrived in my life in the
same year as television graphics

Graphic Design in Television

Douglas Merritt

Focal Press
An imprint of Butterworth-Heinemann Ltd
Linacre House, Jordan Hill, Oxford OX2 8DP

ℛ A member of the Reed Elsevier plc group

OXFORD   LONDON   BOSTON
MUNICH   NEW DELHI   SINGAPORE   SYDNEY
TOKYO   TORONTO   WELLINGTON

First published 1993
Reprinted 1995

British Library Cataloguing in Publication Data
Merritt, Douglas
  Graphic Design in Television
  I. Title
  741.6

ISBN 0 240 51326 6

Library of Congress Cataloging in Publication Data
Merritt, Douglas.
  Graphic design in television/Douglas Merritt.
  p.   cm.
  Includes index.
ISBN 0 240 51326 6
  1. Television graphics. I. Title.
  PN1992.8.G7M46 1993   93–17390
  741.6–dc20   CIP

Composition by Genesis Typesetting, Laser Quay,
Rochester, Kent
Printed and bound in Great Britain

Contents

Preface

The late 1980s and the early 90s have been times of strange contradictions in the television industry. These have been as clearly marked in the working environment of those who produce graphic design for the small screen as in all other branches of broadcasting.

Progress in techniques has reached a peak just when the framework is being radically re-constructed. The pool of talent has never been greater; stable opportunities are at their most confused since television began.

Technology – *the way things can be done* – has had two decades of unprecedented development where electronic and computer sciences have transformed video and film production. The long period of jostling for supremacy between one system and another has been largely resolved. Designers and programme-makers have now absorbed the varied methods and equipment. A sense of stability pervades the marriage of technology and design. Techniques are being used with finesse and appropriateness. Confidence in this area is high.

In violent contrast, the fabric of broadcasting – *the way things are run* – is full of doubt. After years of protection and confidence there is uncertainty, and no section remains unaffected. 'The latest political juggling with the broadcasting system has certainly destabilised the old system.' This bold confirmation comes from the Patron of the Royal Television Society, HRH The Duke of Kent.

Opportunities for prospective graphic designers increased immensely as television transmission and production output grew from the early days of the monopoly of the single-channel BBC. This was followed by the dual-channel environment of the ITV years, extended later by the start of BBC2. Channel 4 brought further opportunities for graphic design, and its unique way of operating fostered many of the independent production companies, who in turn employed graphic designers as freelance contributors. Another expansion was the setting up of the satellite stations; these started with two contenders and were reduced to the present Sky Channel. Now there is the anticipated launch of Channel 5 by 1994 which could add fresh opportunities if the service manages to establish itself.

The development of training has tried to keep pace. Many organisations have been, and still are, deeply involved – outside the efforts of the broadcasters themselves – the BKSTS, The Royal Television Society, CITE, BECTU (the television industry trade union – formerly the ACTT) which has a number of 'accredited colleges', and BAFTA.

Until 1955, with only one transmission from the BBC in this country, there was little sign that design was regarded as a *necessity* in the new medium. Nor was there the clear vision to see that design (not merely graphics) should become an essential factor rather than an optional 'artistic' activity.

The development of design in all aspects of television transmission – titles, programme content, and on-screen promotions – was very slow. Some increase in recognition was prompted in the mid-1950s by the start of commercial television and the mushrooming audience levels. The influence of television commercials, where there were higher production values and an urgency to communicate, were certainly a factor in generating greater design consideration in programme titles. Whatever technical changes occur in the realm of television broadcasting graphic designers will always have to adapt to new methods.

Immense changes have already begun in the way television programmes are made in Europe. The early 1990s will see many more. The very high standards, for which British television is renowned worldwide – in particular for its achievements in graphic design – will remain only if there is a healthy environment in which creative people can work.

Douglas Merritt   Clifton, Bristol   1993

Introduction

Small screen – small world

Designing for the most powerful medium of mass communication yet invented has been a privilege for only a very few people. Audiences are vast but the number of groups originating programmes, and therefore those engaged as graphic designers in broadcast television, has been extremely small. An estimate would be less than 400 in the whole country at the height of activity in the past few years. The total number of art directors and designers working predominantly on television commercials through the advertising agencies would be even less.

The UK inherits the oldest national television broadcasting system in the world – the BBC service commenced in 1936. In 1991 there were an estimated 21.5 million households in the UK with television receivers. Over 95 per cent of the population views television for some time in each week and our average viewing hours in 1991 were 23.8, in the USA it was 49.35 hours (figures from *Screen Digest* February 1992). A majority of countries throughout the world have both national and private television stations and there has been a great momentum towards further channels during the past five years.

Television, as film before it, combines visible and audible messages and it is this combination which makes it so compulsive by bringing the audience as close as is technically practical, at present, to reality – engaging two of our five senses, sight and hearing.

Graphic strength

'Vividly descriptive' is one of the several dictionary definitions of the word 'graphic'. This underlies the main purpose in applying time, money and creative energy to the peculiar art of graphic design. Binding images with sound, in a time base, is the absolute essence of graphic design for television, and those who realise and apply this with imagination and flair contribute greatly to make the whole medium come to life.

The short span of time in which so many graphic messages appear on-screen is often inversely proportional to the amount of time and effort required to prepare the work. Even with the support of electronic and computer-based equipment (now expected by audiences and designers alike) the number of hours required to produce only a few seconds of screen time of moving images, or lettering, synchronised to the soundtrack, are very high indeed.

Effective – but expensive

For every second of film animation 24 frames of perfect images are required (in video, in the PAL system, it is 25 frames) so it is not therefore surprising that a film rostrum camera operator in the past, working with an assistant, could have shot a mere 10 seconds of animation, depending on its complexity, in a day.

A widely held misconception suggests that computers have speeded up the production time to a point where there are very few problems. This is rarely true. Computers have often seduced designers to attempt more complex tasks which take as long, if not longer, than previous methods. The capital cost of equipment, the hourly hire rates, the number of people with many different skills, required to make a few seconds of electronic-based animation can be very high indeed.

In contributing about 30 minutes of computer animation to six one-hour episodes for a medical science series the graphic designers of NHK, Tokyo, needed a very large budget, an in-house team, plus two production houses outside the company, all of them working on the project for almost a year. The background to this venture is given in a case study called 'Dissecting a medical series' in Chapter 7.

A transient medium

Impermanence and restlessness haunt those engaged in this particular branch of graphic art. The final state of their finished work, although visual, often exists only in

an all too complex business. To avoid getting lost in all kinds of complications, many researchers turn to some primitive indicators as well as overall summated scales with blind unidimensional treatments, although these can hardly be considered good or legitimated approaches to multidimensional scaling.

In terms of multi-item scaling technology, modern psychometrics deserves recognition for clarifying the issue of dimensionality and introducing powerful statistical tools for dimensional analysis, especially the popular factor analysis. Nevertheless, despite its sophistication with the use of advanced analytic procedures, psychometrics has stressed the requirement for unidimensionality in scale development. Scaling concepts and techniques, such as the idea of internal consistency, have been developed mostly to address unidimensional scaling needs. The limitation of psychometrics in dealing with extracted, especially un-correlated, factors/components has left multidimensional scaling methods largely incomplete. Since factor analysis is intimately related to issues concerning scientific generalization (Nunnally, 1978), this has set an invisible limit for psychosocial research. Few, if any, notice that theories have quietly suffered from the splitting of different dimensions into different constructs. Indeed, if factor hypotheses set the extent to which one can generalize results across variables that are given the same name, dimensional analysis would in effect become an in-surmountable barrier to the advancement of knowledge in terms of generalization up to the original theoretical levels or back to the original theoretical constructs. This certainly leads to a gap between theory and research.

With the consequence that practice has been poorly directed, psychosocial researchers have been largely left unguided in dealing with extracted factors/ components. There are many people who have endeavored to develop various scales and scaling procedures around important psychosocial constructs. Yet few, if any, have fully proved the utility of the invented devices and explained why. Although an original theoretical idea could be considered extremely important and insightful, it would rarely be deemed as scalable unless it breaks up theoretically or empirically, which, however, would at once wipe out the original idea itself as a general theoretical construct. In the field of social support, for example, Vaux (1992, p.194) asserts that "Social support, per se, is simply not a viable theoretical construct and it cannot be measured."

Such a stance, however, is not substantiated by scrutinizing all the possibilities in measurement. In real terms, the combination of the subscales or different fac-tors often makes good sense in theory, research, and practice. The question is thus not whether a general scale is useful, but how such a scale can be constructed in

the format of millions of stored digital signals on video tape or disc that appear before the eyes and ears of millions of people momentarily and then the work is gone. 'Television gobbles you up' are words vividly recalled from a graphic designer of 20 years' experience.

Memorable single images abound in the world of two-dimensional print graphics created by designers and illustrators like Savignac, André Francois, Paul Rand, Abram Games, Alan Fletcher and other inventive manipulators of the graphic icon. Such strong images are much more difficult to retain with the perpetual movement of television design.

It is possible to recall with pleasure the opening titles of a particular programme a long time after it has been transmitted – *The South Bank Show* for example. But what elements and images are actually remembered? In one version of *The South Bank Show*, lasting no more than 40 seconds, there were at least 20 split-second glimpses of subjects 'plundered' as the designer confessed, from Leonardo da Vinci, Rembrandt, Hokusai, plus drawings of Dylan Thomas, Bob Dylan, Elvis Presley and others, ending on a detail of the hands of God and Adam from the Sistine Chapel ceiling.

When divorced from the movement and sound, and translated to printed stills the graphic images of television lose their potency. They only exist effectively within the context of transmission.

Box of tricks

All visual art has at its base some aspect of illusion. In graphic design for television this element is very prominent. In the early 1980s Thames Television produced an exhibition entitled 'Box of Tricks' which showed the work of the graphic designers. It demonstrated that knowledge of the possibilities for creating illusions is central to much design for television.

There is the same exploitation of 'trickery' in the films made by Georges Méliès using double exposure, (the first shot in 1896, only a year after the Lumiére brothers' presentation of projected moving film), as in a recent television commercial which uses digital electronics when a red silk cloth is pulled, with great panache, from a new car in a showroom to reveal – nothing! The double exposure, the mix, the fade – and now digital editing of video with chromakey. Plus ça change! To some extent communicating through the moving image has changed very little. Advances in technology have given us colour and the 'seamlessness' which we all take for granted. A French television

Still from the British Film Institute

Georges Méliès exploited tricks which he saw in the newly-invented moving-camera. In Indiarubber Head *(1901) he used double exposure and a long zoom to enlarge his own head then he appeared in the scene. Almost a century later the Video Paintbrush Co. of Sydney make similar tricks with Harry, very simply*

This symbol, itself an optical illusion, was designed for a Thames Television exhibition to emphasise the 'tricks' inherent in TV graphics. (Below) Computer technology achieves combinations of multi-plane moving images in ways and at speeds that could not be achieved a decade ago. Barry O'Riordan, Art Director at MPC, designed this 30 second-title sequence for Artworld *for the producer Stuart Binns of Transworld International to portray aspects of art from every part of the world. The computer animator was Gareth Edwards and the 'Harry/Paintbox' work by Rob Hodgson, both of The Moving Picture Company*

▶ Throughout the book this symbol denotes the beginning of an animated sequence

▷ This mark indicates the direction the sequence follows

■ A black square appears at the end of each sequence

programme in homage to Méliès made the comment 'Duplex photography, superimposition and substitution. Have these been improved since? No. We just have better equipment!'

That audiences of millions can now share live moving pictures, with sound, via satellite from anywhere in the world at the same moment, at the touch of a button, and record it themselves, does not seem strange to the present generation. A teasing cartoon from *Punch* in 1878 made the then totally improbable prophecy that a rich household might, one day, be able to 'tune-in' to opera from various theatres throughout the world. By 1907 this prophecy was considered fulfilled with the introduction of the gramophone. Selecting from multi-channel transmission, listening to stereo sound, sharp images in full-colour, recorded if you wish while you are watching another programme, or even many miles from your home, might have surprised even the most audacious Jules Verne.

Creative prospects

Whatever technical changes are made there is a limitless world of creativity for future generations of designers. They will investigate and exploit the moving image in their own way.

The intention in these pages is to look at the following: a picture of the environment in which television graphic designers are likely to be working in the near future; some of the attributes and skills they may require; the education and later the training they will undergo if they are to become involved in this medium; a guide to the mass of equipment now at hand; and the paramount importance of *imagination*. 'The mental faculty to form *images* or concepts of objects not in existence' according to one Oxford Dictionary.

Without any doubt, harnessing the benefits of the microprocessor to graphic design in television has resulted in a far greater volume of graphic work on-screen every minute of the day or night in all stations throughout the world. Designers themselves have gained more prominence within the profession. The digital graphic workshop has become a reality; the image-producing work station at the graphic designer's desk is an ever more likely possibility. The potential to deliver design effectively is now much greater.

Irrespective of technical advances there will always be as many different kinds of designer as there are people. No single role model or set of qualities can ever be prescribed. □

Musical Mistress of House ("on hospitable thoughts intent"). "Now, RECOLLECT, ROBERT, AT A QUARTER TO NINE TURN ON 'VOI CHE SAPETE' FROM COVENT GARDEN; AT TEN LET IN THE STRINGED QUARTETTE FROM ST. JAMES'S HALL; AND AT ELEVEN TURN THE LAST QUARTETTE FROM 'RIGOLETTO' FULL ON. BUT MIND YOU CLOSE ONE TAP BEFORE OPENING THE OTHER!" Buttons. "YES, MUM!"

This 1878 Punch *cartoon foresaw the broadcasting of music to the home. Thirty years on the prophecy was considered fulfilled with the gramophone's technology. What changes will we see in television programme-making, transmission and interactive and multiple channels to revolutionise graphics?*

Time Present.—MUSICAL MISTRESS OF HOUSE: "NOW, ROBERT, BRING ME PATTI'S GRAMOPHONE RECORD 'VOI CHE SAPETE,' ALSO THE QUARTETTE FROM 'RIGOLETTO,' SUNG BY CARUSO, SCOTTI, MISS ABBOTT AND MADAME HOMER, THEN GET ME THE LAST VIOLIN RECORD BY MISCHA ELMAN, AND DO NOT FORGET A BOX OF MELBA NEEDLES!" BUTTONS: "YES, MUM!"

Chapter 1   The Ecology of Television Graphics

Interference with the picture

Many changes in the broadcasting structures in the UK have occurred – others seem inevitable. 'All four channels have shared the same broadcasting ecology. All have benefited from the stimuli, innovations and competitive professional pride of the others. That ecology is now at risk.' Peter Kellner wrote this in an article in *The Independent* (18 October 1991) on preserving the quality of broadcasting. The old system was protected but it provided standards that, literally, 'money cannot buy'.

For graphic designers, and many other creative crafts people, television provided opportunities to work in a very wide band of experience.

In an introduction to the book *TV, UK*, by Jonathan Davis (published at the end of 1991), the problems of the broadcasting community were described as:

> 1991 has been a year of unprecedented turmoil . . . it began with the fall-out from the BSB/Sky merger and will culminate in the agony and the ecstasy of the Channel 3 licence awards. A fierce economic recession and the collapse in television advertising revenues have brought with them deep soul searching and anxiety about the future.

Graphic design departments in the BBC and in ITV have already, like their parallel services of costume, set design, make-up, lighting, engineering and technical operations, begun to shed staff and have reduced all levels of activity through 'natural wastage', 'early retirement', as well as the more painful expedients; capital equipment purchases of new and replacement items have been postponed or abandoned.

Recruitment of newcomers has slowed almost to zero. The most recently appointed designer in the Graphic Department at BBC Television Centre in London during the autumn of 1991 was a very experienced person of more than four years' service.

Training is a core element that has suffered in such conditions. No matter how good the education of art college graduates may be, the only way to become immersed in the highly-specialised environment of television graphic design and the complexities of production is by working alongside those who have absorbed the work experience themselves. This is true of every profession – dramatically displayed in medicine and law.

Grants, sponsorship for studies and scholarships have already begun to dry up (e.g. Thames Television's Travelling Design Bursaries, whereby for nearly 20 years funds were awarded to talented art students, have been suspended). Will the new broadcasters' 'response to market forces' enable them to sustain these or to create similar enterprises?

Graphic design is a service at the command of those who have control over what is made and transmitted. Designers cannot decide the content of what they produce. They can, however, affect the audience's response very strongly.

There is enough evidence to show that the current changes may reduce the output of news, current affairs, programmes on the arts, and documentaries, and may alter the mix of programmes to be made. Television drama has already withered through lack of funds.

Increased competition for the finite amount of advertising revenue and audience figures, from a self-supporting Channel 4, the new Channel 5, cable and satellite stations, plus the growing video recording interest, will make ITV truly commercial for the first time by the mid-90s. If primetime were ever to be filled with game shows, very cheaply made programmes, or more and more repeats, the scope of designers' work would become impoverished, and to hold their 'audience share' the BBC would surely follow.

The 'ecology', evoked by Peter Kellner at the start of this chapter, that 'all four channels have shared', has meant in the past that money has been available to nourish a very diverse mix of material. Could programmes like *World in Action*, *This Week* and *The South Bank Show* be transmitted at peak viewing times when advertisers or sponsors may not be willing to support them? Large sums of money have been devoted to schools' broadcasts and other educational programmes by the BBC and ITV in the past where the profit motive has not been the primary force.

The electronic generation of both still and animated images has made working with paints and paper less and less familiar – a loss regretted by many

There has been undeniable richness of material for many decades and part of the fabric has been the enthusiasm for graphic presentation engendered by programme controllers, producers, directors and editorial staff in every type of programme. Without educated 'patronage' little of value occurs.

In an overcrowded arena much young talent is certain to remain uninvolved while those in mid-career have been thrown off course. For large numbers this will be seen as a period of painful change.

Many of the designers who are now impelled to leave their jobs in the industry believe, as one designer said, 'the readiness of managements at the BBC and ITV to 'squander the wealth of experience acquired by their staff shows that they merely paid lip-service to the value of design'. Peter Le Page, the Controller of Visual Services at Thames Television, said 'There will be an interregnum of lower standards and confusion, with large numbers of people being employed as "casuals", before the third age arises'.

Contrast and brilliance

A result of these major changes in the climate has been the 'freelance explosion'. On the positive side this has promoted the expansion of independent graphic design groups in an industry that had previously used only the most minimal freelance component. Groups began to form outside the BBC and ITV some 10 years ago, on a small scale at first, against considerable odds. Having gained status and produced good work, will there be enough to sustain freelance groups and individual designers in the industry they hope to serve during the next few years?

They were encouraged by the patronage of Channel 4 where enlightened use of graphic design was seen as being an integral component within the channel's programme making and self-promotion.

The high proportion of BBC-trained designers among this 'happy band of brothers' and sisters is noteworthy. They are well placed to work for their 'old masters' – ITV, Channel 4 (and maybe Channel 5), the international satellite broadcasters, as well as for the independent production companies who are eager to find opportunities in the more open market. If the programme controllers get their calculations right in holding audiences, and the advertising revenue sustains the expanded industry, their future looks very good. Commissions from abroad, covering programme graphics, station idents and art direction for television and cinema commercials are already part of the graphic entrepreneurs' success. Their effect on all television design has been profound and will prove lasting.

Will 'normal service' be resumed?

In the past the rivalry between two protected systems (BBC and ITV) for the same audience resulted in high standards in all aspects of making programmes.

When this formula is disturbed these aims may be more difficult to maintain in the 'commercial' arena than anyone anticipates. Those who sought the changes may regret the change of diet – reversal will be difficult.

Graphic design is one of the elements that can be cut back very easily. It has often been marginalised.

'Sets?' 'They are essential.' 'Catering on location?' 'Can't do without that!' 'New titles?' 'No. Let's wait until the next series . . . or maybe the next.' □

Chapter 2   Framework for Television Graphics

There are nearly a dozen different circumstances in which a graphic designer might be employed in television in the 1990s. What are these circumstances and how do they they differ?

The first, and longest established possibility, is as a designer in a *graphic department at the BBC*. A second is as a permanent *staff member of any of the ITV contractors* – this could include 'GMTV', the new ITV breakfast contractor. Thirdly a graphic designer could work for the presentation department of a *'publishing contractor'* like Channel 4 or Carlton Television. A fourth possibility is working with a *satellite channel* or *cable station*. A fifth is working with one of the *independent production companies* that emerged at the formation of Channel 4 and that have proliferated with the policy of deregulation. A sixth is employment within one of the ever-increasing groups of *private graphic design groups*. A seventh is as a graphic designer or art director, at a *post-production, television facility house or computer animation company*. Working for a news *channel*, (e.g. Channel 4 News) is an eighth option. Working for an *advertising agency* where television commercials are main-stream is a ninth option. A tenth alternative would be as a *sole freelance*, when a designer can work for any of the others as and when required. An eleventh could appear with the arrival of *Channel 5*, if and when it takes off.

Employment framework up to 1980

Designing for television graphics from outside the departments of the BBC or any one of the 15 ITV companies was unheard of prior to the early 1980s.

A self-contained internal group at the BBC grew as the service demands of BBC1 and BBC2 were increased by longer transmission hours, more programmes with more elaborate graphic presentation, and then by the additional work demanded by the introduction of colour in 1968. But the corporation rarely employed freelance graphic designers, although certain services, like cel animation, model-making and rostrum camera shooting were contracted from outside suppliers as required.

ITV did not seek to use outside designers. The agreement between the companies and the union (ACTT at that time) virtually forbade the use of those who were not members. Only a handful of graphic designers who chose to work casually, or on short-term contracts, moved from one station to another.

In graphic design for print, publishing and exhibitions – where the structure of the industry was established well before World War Two and where there were no trade restrictions – many private design groups changed the momentum of British design as they flourished from the mid-1950s onwards.

In television everything has changed dramatically in less than 10 years. Independent graphic design groups now proliferate, but for a very long time there were only the two options – BBC or ITV.

The current and future framework

Staff levels, the type of equipment, the working conditions and the design ethos vary considerably in these organisations and profiles of some of them are presented below. Government changes have largely removed the opportunities of permanent employment and training the BBC and ITV sectors.

1 Graphic design with the BBC

The British Broadcasting Corporation has the longest established public service television transmission in the world. It was started in 1936.

The influence of the Graphic Department at the BBC under various heads of the group cannot be over-estimated. The department can be seen to have set the standards in creative design, in aesthetic treatment and in developing technical methods since its inception. As the first graphic designers in a new industry with a privileged non-commercial ethos, all members of the department have tried to retain this special position.

It appears from this photograph that Her Majesty the Queen was made aware of the importance of 'graphics' when she made a tour of the BBC with the Director General, Hugh Carlton-Greene, in 1955. She may have witnessed the production of the archaic hand-set and hand-proofed caption cards for an election programme. (Right) The results are being hand-written! The general view of the graphic design area at that period does not suggest great creative verve

However, the title stills opposite give a taste of the high design qualities in the days of 405 line and black-and-white transmission. Detective *was an early design by Bernard Lodge when at the BBC and John Tribe designed* Division *for the old ITV company Associated Rediffusion*

Recruitment and general management of the graphic design services in the many regional BBC locations – Bristol, Manchester, Birmingham, Leeds, Edinburgh and other provincial centres – has always been overseen, if not directly controlled, from Television Centre in London. This has done much to maintain standards of technique and a conformity of style. Those well trained in the past at the BBC have spread their energy and skills in every possible branch of television graphic production.

A compilation showreel of titles made by the BBC, with material from 1955 to the present day, gives a mere hint of the range of the subjects, the ingenuity, the constantly changing styles and the techniques used by generations of designers in their graphic design work. *Dr. Who*, by Bernard Lodge, *Famous Gossips* by Alan Jeapes, *I, Claudius* by Dick Bailey, *Tomorrow's World* by Pauline Carter, *Soldiers* by Liz Friedman, stand out after more than an hour's viewing. Only a minute fraction of the massive outpouring has ever been recorded for archive purposes.

A general concern is that this achievement may not be possible in a future where a quota of 25 per cent of the annual production *must* be contracted to outside companies. This is heightened by the decision that, in common with other BBC services, the graphic design department will in future operate under what is known as 'producer choice'. In this new formula the programme-making producers are now free to purchase the creative graphic design and production from *any source* they wish – so long as costs are within their overall budget. This system means that from a total fixed sum that the producer will have contracted to supply a single programme, or series, the amount left for 'graphics' will often be set *without any relation to the concept, storyboard or modest production costs*. By tradition the graphic designer is responsible for the production of *all* the work. The idea that they produce ideas and storyboards for others to translate for screen has never been true.

Another change in the 'rules' will mean that the 165 members of the BBC Graphic Design Department will now be able to undertake outside commissions from anywhere in the world. Already work is in hand, or being negotiated, from Austria and Germany. A helicopter pad is probably already on the roof at Television Centre, waiting to fly designers to Heathrow! A Marketing Manager has been appointed to persuade producers not to desert the in-house design team, and presumably to encourage everyone else in the world to 'Buy British Broadcasting'.

With wide deregulation, events have moved fast. Many contracts for BBC graphic programme titles and other work have now been awarded directly, or through independent production companies, to graphic designers employed by ITV companies! Two graphic designers at Thames Television have gained a contract to design the titles of the BBC2 series *Heart of the Matter* for Roger Bolton Productions Limited when three other design groups 'pitched' for the work. (A report on this

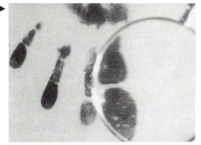
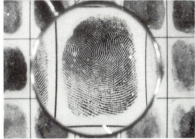

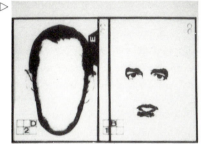
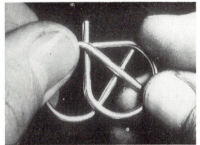

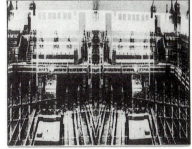

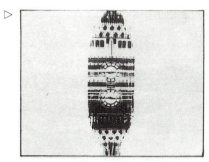

is given in Chapter 7.) Will there be more pitching than tossing? The potential for Byzantine relationships in the future seems limitless with ITV designers commissioned by independent producers for BBC programmes or BBC designers commissioned for ITV productions!

2 Graphic design in ITV

From 1955 the ITV companies were set up on a regional basis. Each employed graphic designers in-house, and for many years the five largest companies (Thames, LWT, Granada, Central and Yorkshire) shared the job of producing on-screen presentation material for the whole network.

The range of subject matter and quality of programme work for graphic designers in ITV have broadly matched those of the BBC and the commercial system has provided opportunities to design for large scale documentary series like *The World at War* and *Hollywood*; and feature programmes like *World in Action*, a Granada programme still in production, and *What the Papers Say* one of the longest running television series of all. The list of ITV drama programmes is distinguished and long. Many people have described the period from the late 1960s to the mid-1980s as the 'golden years' for British television.

The ITV graphic groups varied from about 25 to one or two designers, depending on the size of each station. They were backed by stills photographers, phototechnicians, typesetting facilities and, in some companies, rostrum camera units. In more recent times they acquired character generator operators. Some graphic designers made very close links with engineering and the technical operations departments, finding new ways to achieve the transferring, editing and recording of graphic material. Among these were the earliest attempts to produce video animation from cels and simple flat artwork – made by recording on to tape with a VTR machine – long before video rostrum cameras or stills stores were available. A list of the current Independent Television Contractors in the UK is given in the Addresses section of this book.

3 Working for television 'publishers'

From January 1993 those awarded franchises as contractors to serve the regions in the UK will continue in much the same way as the system has operated since 1955, although some companies will not make their own programmes. Carlton Television in London is, for example, to be a publisher/contractor. Like Channel 4 before it, Carlton will *only commission* new programmes. In addition they will buy films and other programme material from any source they wish. This means that companies like Carlton will not employ graphic designers directly, except to service their own on-screen presentations.

Tim Simmons is the Head of Presentation and Promotion at Carlton and the design department working under him in-house will be led by a design manager and consist of two strands. There will be two designers (one Senior Graphic Designer, and one Graphic Designer) responsible for the on-air graphics to establish the new Carlton corporate image (Lambie-Nairn and Co have designed the Carlton on-screen ident). The second strand, of two picture co-ordinators will research and produce the on-air stills using a Quantel electronic paint system and a stills store.

Simmons' plan is for both strands to combine and achieve a very close liaison in the design of Carlton's printed publicity using Apple Macintosh for print and on-screen work – an area few television companies have managed to handle – with the strong overall corporate appearance he wishes to convey.

The early days for the new company may be complicated. Carlton Television will wish to be seen as part of ITV, yet establish its own new identity at a time when the older companies are merging theirs. Solving such problems is the essence of good design management.

4 The satellite sector

Sky Channel, as it began (the channel is now called 'British Sky Broadcasting'), was the first opportunity for British graphic designers to service satellite television when it went on-air in February 1989.

Satellite and pay cable are the fastest growing sectors of television reception in the UK, increasing from an annual turnover of £18 million in 1986 to £262 million in 1991 (Broadcasters' Audience Research Board).

In a new building on a business park in west London, where staff and equipment were swiftly installed, they now produce graphic work for every aspect of on-screen presentation for the massive output of all six channels. These comprise Sky One, The Movie Channel, Sky Movie Plus, The Comedy Channel, Sky News and Sky Sport. Four of these have 24-hour transmissions. The Comedy Channel runs a mere seven hours a day but Sky Sport is usually active for 21 hours.

The whole graphic group of 40 full-time in-house

Carlton Television's on-screen ident was designed by Lambie-Nairn and Company. The top three stills are from on-screen promotion work by that company. The two on the left (bottom row) are by the chief in-house designer Dean Stockon and one on the (right) is by staff designer Jackie Martin

people is lead by Mike Hurst (ex BBC, ex TVam and ex ITN). He has four Graphic Design Supervisors, 10 Graphic Designers, nine Assistant Graphic Designers, two computer programmers and 18 graphic technicians for 'capgen' and electronic library systems.

Taking off, if that is what satellite channels did, in the late 1980s, Sky recognised the need for electronic graphic production and provided this from the outset. In three video graphic suites there are five Quantel Paintboxes: three are 'V' series, of which one is linked to a Harriet and one is dedicated to the Sports Channel, and two Classics, one of which is dedicated to the News Channel. In digital effects (DVE) equipment Sky has two Abekas (an A53 and an A64 with a 601 edit controller using D1 format). In-house three-dimensional computer graphics share two Symbolics machines and a Silicon Iris.

Hurst expects more animation will come from the Apple Mac equipment, as techniques advance. At present this is used to generate artwork and type effects which are beyond the scope of the caption generators.

These hardware emplacements mean that there is very little need to travel from west London to the West End facility and post-production companies. With such

production there is little opportunity to do so. The large workload embraces:

(a) Developing and maintaining a housestyle for each Sky Channel. (Five of the original on-screen idents were designed and produced in the USA by the designer Harry Marks of Los Angeles to create a 'transatlantic non-BBC look' for the whole enterprise).

(b) Title sequences for programmes and many of the films.

(c) Channel and cross-channel promotion animations.

(d) Major promotions for main series, e.g. announcing details of the *Premiere League* for Sky Sport – *Monday Night Match, Premiere League News, Rugby Information* and *Premier Movies*.

(e) The routine but essential work of 'run-downs' or 'menus' for forthcoming programmes.

(f) The design group supplies graphics for 30-minute news bulletins on the hour, every hour, night and day in a format similar to CNN (Cable News Network in Atlanta). The remaining half-hour is commentary on world news and that too has large graphic input.

a legitimate manner, or with approaches that are at least theoretically and empirically plausible. This is a fundamental issue in psychosocial scaling. In sociology, for instance, there has been a lasting debate as to whether the discipline should pursue grand, piece-meal, or "middle range" theory. The issue has a great deal to do with the problem of generalization. There are also heated talks about the link between micro and macro social research. However, the discussion will lead to nowhere if this kind of "generalizability" problem remains unresolved.

Those who have tried to construct a global scale by combining different sub-scales or original items have often been trapped in a theoretical pitfall, however. They have somehow related factors that do not show meaningful relationships, or in a way that does not reflect any reasonable associations. It is, indeed, all too arbitrary: There is no rationale seen in this regard that would justify the seeming violation of mathematical laws. This is, of course, a practical issue; its solution, however, demands some sort of theorization. As is the case in science, most of the theoretical stipulations should eventually take into account the complications of the real world and the needs of practice.

Unsatisfied with the way theories have been represented in psychosocial research, Chen (1997) sets forth a central theme of unidimensionalization and articulates some theoretical and practical approaches to its improvement. Here unidimensionalization means the *combination of measurement items into a single scale by relating the different dimensions of a data set through theoretical and/or empirical means*. Conventional multidimensional scaling including factor analysis can only deal with the issue of multidimensionality by extracting a certain number of factors or components. Subscales of a multidimensional construct can be formulated based on such results. The issue of how to combine these factors or components, i.e., the problem of ultimate unidimensionalization, however, demands a thorough theoretical understanding and a clear research purpose. Scientific facts can easily be distorted by blending different dimensions without pondering their relationships. A review of current literature indicates that the practical strategy in this regard is generally weak and inadequate: Either the opportunity for further unidimensionalization is simply given up, or it is accomplished in a way that is not fully justified. In other cases, factor analysis or any other multidimensional scaling procedure is not performed, which leaves the scales far from established. The reason is straightforward: The goodness of fit of a dimensionality model accounts for a primary source of errors.

According to Chen (ibid.), unidimensionalization is central to multidimension-

This volume of work is impressive. All the graphic work is allocated and supervised by the Head of Graphic Design and presented to the head of each channel. Promotion work is 'commissioned' and approved by the Head of Presentation.

To achieve this output all the graphic staff, designers and operators, work on shifts and they rotate from one channel to another. This gives them a variety of styles and exposes them all to the detailed needs of every channel's production system.

At present there are 17 digital library stores which 'bulge' with still images. On-screen type generation originates from 14 Chyrons and two Aston Caption machines, one with Wallet. Two of these Chyrons are 'masters' in one of the video graphic suites. The remainder are booked and shared as required.

Displaying the sponsors' idents in animations and stills, to stop them looking like inappropriate tattoos, is a growing graphic design occupation.

Mike Hurst says, 'It's good for designers to adjust to many modes. With six channels the variety of subjects, and discipline of 24-hour news, we present as much of a challenge to designers as the whole of BBC Television Centre.' He welcomes the strong and demanding interest of Gary Davey, Sky's Deputy Managing Director, in the continuous generation of BSB's graphic pulse.

5 The Channel 4 work expansion

New ways of working were created by Channel 4's remit to *commission* all its programmes – not to make them. This created possibilities for independent television programme-makers as the channel does not have to employ creative staff, other than a very small design and art direction service in-house. Only two or three designers have been needed to handle the channel's on-screen promotions work, always known as 'Presentation' (or 'Pres') to television staff, and supervise the 'house-style'. In the first year Channel 4 had over £90 million to spend on programme commissions. The emerging groups of television graphic designers now had five markets – BBC, ITV, Channel 4, television commercials and pop-videos, where some very unusual, often expensive, work was commissioned.

Channel 4's success consolidated the position of the independent programme-makers and the spread of the share of work for the emerging 'free' designers.

Grouping of like-minded and similarly-trained designers proliferated, and many designers foresaw the break-up of broadcasting monoliths as an opportunity to start their own companies.

6 Employment with independent design groups

The first TV escapologists of note came from both camps. Two ITV graphic designers at London Weekend Television decided they were dissatisfied with the prospects and status in their full-time posts. Colin Robinson and Martin Lambie-Nairn resigned and set-up their own private company, designing for print and TV, gaining commissions from LWT while retaining their union membership. They were commissioned to produce graphic information animations for the LWT Sunday programme *Weekend World* . The company, Robinson Lambie-Nairn (now Lambie-Nairn & Company) prospered and the ratio of television to print work increased, reaching a peak when they enterprisingly gained the commission to create the on-screen ident for Channel 4, setting that newly-formed broadcaster in a strong design direction. Jeremy Isaacs, the first chief executive of Channel 4, was quoted by *Broadcast* with the compliment, 'Graphics are the thing that people know us by and we are the envy of the world for them'.

Bernard Lodge, the graphic pioneer of *Dr Who* title fame, left the BBC after many years, drawn away partly by his experimental film rostrum camera work with Filmfex. As he gained work Colin Cheesman, then Head of Graphics at the BBC, joined him as a partner, and as Lodge/Cheesman they added to the climate of change and the entrenchment of independent groups by concentrating solely on design for television graphics.

A few independent designers were soon invited to work for foreign television stations. British design groups are now sought by broadcasters and advertising agencies throughout the world. Their reputation for creativity and technical acumen is internationally accepted to be among the very highest. They seem to share, with British pop-groups, the art of making a product that 'travels well'.

Prestigious foreign contracts are exemplified by the recent award of the whole of the on-screen identity for the first commercial terrestrial television station in Norway, TV2, to the British independent company Plume Partners. Plume's founder Richard Morrison and his partner Paul Blake, who directed the Norwegian project, have built up a portfolio of television design and cinema work since Plume Design started in 1970 and Plume Partners was formed in 1989. When at the BBC

Paul Blake worked on the launch of *Breakfast TV,
Panorama* and *The Money Programme.*

Something about the nature of graphic design for the
small screen inhibits the one-man band. Designing and
producing film and video animation is complex. The
variety of production avenues seems endless and it is
this aspect which demands the support of a number of
people working together. New roles have emerged. One
of the most important is that of the 'producer'. Here
placing and overseeing the work from design to
completion for delivery, including the budgeting, is a
separate responsibility. Costs can be considerable and
production times can be months rather than weeks.
Good management for the sake of both the clients and
the designers is imperative. An experienced producer
describes her work in Chapter 10.

In spite of 'being a group' each designer usually works
creatively alone. They may co-operate at other levels but
each commission is almost invariably handled as the
creative work of one designer – senior or junior.

Many of the group names echo a desire for the
professional status of accountants and solicitors. 'English Markell Pockett', 'McCallum Kennedy D'Auria',
'Baxter Hobbins Sides' and many others followed where
'Lodge/Cheesman' and 'Robinson Lambie-Nairn' had
led. The group named 'Blink' is a notable exception.

Whatever they choose to be called, these companies
have been a strong creative force and have improved
the standards in programme making and TV commercials. The following reviews of a selection of the
'independents' will show their valuable addition to
graphic design for television.

Ortmans Young

Both partners in this design company served the classic
apprenticeship at the BBC and at Channel 4 before
deciding to set up their own in 1987. Marc Ortmans
bases his reasons on this hypothesis: 'Designers need
their freedom. They want to be loved and respected by
their clients. They invariably prefer and thrive on a mixed
diet of work.' Breaking free provided these elements.

Like many of their design contemporaries, who broke
away at about the same time, they felt they knew little of
business practice but realised the benefit of listening to
the experience of advisers.

With an extensive track record in broadcast graphic
design Marc Ortmans and Haydon Young set out to
broaden what they did and make their debuts in the
allied fields of advertising and corporate communica-

*Stills from Plume Partners' animated ident designs for the
Norwegian channel TV2. The independent graphic design
groups in Britain have gained many overseas
commissions and Plume have now gained the Ulster
Television corporate identity work*

tions. From their Soho offices they have, since 1988,
gained an international reputation, broadened their
client base and met the challenges of designing for
other cultural values, whether they were American or
Arabic, testing their design ability and intuition.

Their compact size has not prevented investors from
knocking on their door, and they have already declined
four offers to buy the company. 'Freedom', Ortmans
repeats, 'is worth more than money'.

They are looking to design and direct more commercials as they feel these provide the opportunities to
achieve levels of creative and technical excellence
greater than the lower budgeted broadcast projects.
Unlike traditional production companies, Ortmans

Young like to be involved as early as possible in every project. This gives them the opportunity to shape an idea before it becomes too rigid, or has its full potential lost to client politics and schedules. An account of their work on a television commercial, '2010', made for the Midland Bank, is given in Chapter 7.

McCallum Kennedy D'Auria

All three design partners left the BBC at the beginning of 1989. Like most groups they were sustained in the beginning by work from the 'Beeb' but within six months only 25 per cent of MKD's design commissions were from the corporation. ITV contractors and many advertising agencies were seeking their very individual talents.

Graham McCallum's work at the BBC from the late 1960s encompassed design for children's programmes, where his ability as an illustrator led him to direct 50 animated films. As book illustrator he has 26 children's books to his credit. After two years' work in the BBC Presentation Department he moved in 1976 to the main graphics area to design many programme titles. Two major series followed – *In the Looking Glass* with the Wells, Bird and Fortune team and *Jane*. Both of these gained BAFTA prizes for graphic design.

He worked very closely in the BBC's development of an early digital painting device (code-named Eric, and later marketed by Logica under the name 'Flair').

In Canada he worked for CBC on *The King of Friday Night*. This won several major US and Canadian awards. He returned to the BBC before he met and joined John Blair, who was then a freelance television director and who then became the Managing Director of MKD.

John Kennedy left St. Martin's School of Art in 1980, worked as a freelance designer/illustrator for two years, joined the BBC as an assistant in the Presentation Group and then collaborated on the innovative electronics-based *Jane* series with McCallum. In 1989 he won a Royal Television Society award for his title for the documentary series *Inside Story* which used Quantel's 'Harry' to weave together film negative and lens effects.

The third design partner, *Paul D'Auria*, completed an MA in Information Design and entered the world of television by joining TSI Video, a post-production company, where he learnt to operate electronic paint systems and other video graphic equipment. With this experience he was approached by the BBC and invited to assist with the introduction of the all-electronic *Six*

O'Clock News programme. This led to him becoming a designer on the *Nine O'Clock News* supervising the programme on a daily basis. Anyone who too easily thinks of designers living in 'ivory towers' would blanch at the pressure of work in such environments.

Training senior designers to use this new technology was Paul's next venture before he broadened his own design experience in the general graphic department. Storyboarding for programme titles and content work involved much longer periods of concentration than the night-by-night news commitments. Like other designers, he worked on a variety of programmes from the light entertainment of *The Lenny Henry Show* and *French and Saunders* to *Antenna* for science and features. He already has three RTS award nominations and two American International Monitor design awards.

'Blink' is beautiful

A much smaller, and in many ways an 'alternative' design group has been developed by a more recent generation. Its unconstrained approach is typified by Mike Bennion who is one of the designer/directors.

His animation work has been praised in the trade television press as 'refreshing', 'always strong on visual ideas' and containing 'a child-like enthusiasm'.

Bennion credits this attitude to the freedom of his three years at Kingston Polytechnic. This was followed by the Royal College of Art from which he graduated, in 1987, with others who have quickly made their marks in an all too often strait-laced design marketplace. Among these were Andrew Altmann and David Ellis who helped to found the equally provocatively named design group Why Not Associates.

Mike is completely relaxed in his pursuit of every possible production method. He has said, 'I think the most interesting way to go is to take the best technology and battle with it.' This has led him to confront: the Apple Mac for a large part of his work; computer animation for the titles of *DEF II* for the BBC; Harry application for 11 three-and-a-half-second stings for LWT's *Night Network*, and live-action model animation for commercials for Taunton Cider's Piermont apple drink.

He has not been afraid to use parodies and kitsch of the wilder kind. Commissions have arrived from almost all divisions of broadcast TV and from advertising agencies. His wish to challenge the expected was shown when he was invited to contribute to a television graphics symposium organised by Studio Dumbar in

The Hague. He promised his audience on the first night that he would make a 60-second sequence in 24 hours. To this end he video recorded delegates in various movements then flew back to London where Cell Animation had agreed to give him access to Paintbox, Harry and Encore to build up a compilation, free of charge, before he returned to Holland, titles in hand! Lateral thinking of this kind, with implied wit and irreverence, is vital to a design profession that can become too serious.

Blink has given him the freedom he craves 'to do everything himself', and he appears grateful in recognising the opportunities that working with 'Blink' has already given him.

7 Design opportunities in facility companies

Many of the private companies who invested early in the newly invented, and at that time commercially untried, electronic equipment were founded by graphic designers or technicians close to graphic production.

Peter Claridge set up CAL Videographics, and Roger Fickling started Stylus in Cardiff. Both these post-production and computer animation companies were quick to employ graphic-trained staff. Their assets were applied in at least three ways.

With television design experience they could act extremely well as art director/producers to guide clients from storyboard to final computer-based animation. Gareth Edwards, trained as a graphic designer but with an exceptional aptitude for computer concepts, did this very well at CAL, and is now with The Moving Picture Company. The facility companies also offered design services and they took on graphic designers from the BBC and ITV to extend this side of their activities. Among many others Graham Kern joined CAL from the BBC and Barry O'Riordan went to The Moving Picture Company from Thames.

A third function came from those graphic designers who became interested in production services and equipment. Their interests, enhanced by their design backgrounds, brought them into the facility world as *operators* and technicians. Sue Land started as an assistant graphic designer at Thames. There she was among the very first to become adept on 'Paintbox', graduating to carrying out information graphic animations for *This Week* on the Harry device at Cell Animation. Later she moved to Stylus in Cardiff where her understanding of the capabilities of the 'Harry' device was heightened by her design faculties.

8 Design work for News

Most television stations have had to take on the production of news programmes. These are very high in staff levels and in equipment costs. News has been an important catalyst in promoting graphic input and speeding the use of computer-aided design. For many years graphic designers were unable to deliver what was required because of the slowness of production of even the simplest image when relying on hand-work and Letraset and the hope of minutes of animation were a mere day-dream.

The Americans led the way with early digital painting systems like AVA (Ampex Visual Art), and an Aurora was soon in operation at KRON, an NBC affiliated station in San Francisco at the beginning of the 1980s. Following them, CBS News in New York used a stills store, and the news graphic designers at ABC in the same city were making complex animations for their editorial team using a Dubner system. This was a character generator with some 'paint facilities' and real-time animation. In 5 year's time there is unlikely to be a news service without an 'electronic graphic workshop', as they were swiftly designated, and it is difficult to recall the strong criticism of computer graphics on the grounds of poor aesthetics, bad resolution (before antialiasing was introduced), and lack of good type-faces. The first global news-only station at CNN in Atlanta uses graphic design teams 24 hours a day.

Now the BBC, ITN, and the main ITV stations are all equipped with painting systems, character generators, effects devices, and editing systems like Harry and Harriet. But the key ingredient was the large capacity digital photographic library, or stills store. The introduction of this component is viewed in Chapter 8.

9 Graphic design for TV commercials

The terms of contract when employed as a graphic designer for broadcast television, whether for BBC or ITV, usually preclude any work for television commercials. Only when experienced designers broke away from either of these employers in the mid-1970s were they free to design for advertising agencies.

Prior to that time most commercials were directed by those working as film directors. Now they work side by side, merely exploiting their own mores – the graphic designers relying on animation, in all its forms, and the film directors offering their feature film vision.

The emerging independent graphic groups' design work for commercials has extended their output and added to the creative and technical presentation of television advertising. Graphic designers in TV have most clearly a unique knowledge of designing with video and computer effects, born in a decade of close collaboration with facility and post-production houses. Swaying cows, and more recently the mysteriously moving buildings in a Neefax commercial, designed by Ortmans Young, exemplify this welding of design and technology.

For British television graphic designers the prospects of the international advertising market in the next few years appear bountiful.

10 The lone freelance

There is only a small requirement in the television industry for the single freelance in graphic design and these places are more likely to be for technicians and operators rather than designers. For holiday relief and to cover staff in the intense atmosphere of 24-hour shift-work, especially in news and current affairs broad-casting, there always seems be some demand for short-term or 'casual' staff. Many younger designers are entering television by this route and some have opted to remain freelance for many years. Those who can operate anything from a character generator to any digital editing or effects device – and relatively few people can – are eagerly sought. In 1980 the trade journal *Broadcast* headed a feature on the burgeoning television technology 'Will graphic artists take the electronic tablets?'. They did, and for some it was a key to their future career.

11 Prospects in Channel 5?

Tenders for Channel 5 were made by the Independent Television Commission early in 1992. The business hopes for raising capital for the project were not optimistic.

Major problems dull the prospects, and newspaper stories with headings like *Channel 5 Sends Out Weak Signals* set the tone. The greatest problem is the cost of having to convert a large number of VCRs to avoid interference from the new signal. Separate aerials will be necessary for a majority of viewers to this new channel and even with these only 70 per cent of the country will be able to pick up the station.

Vast start-up costs were predicted. Then Silvio Berlusconi, the moghul behind Italy's commercial television, and other bidders withdrew, leaving Channel 5, a consortium led by Thames Television, as the sole

contender. In December 1992 the ITC rejected the terms of the Thames' bid and the opportunity of launching this additional channel for the next century has evaporated.

Inter-continental graphic design

Some graphic designers of the future will be working for very large international companies. They will travel between, or be based temporarily in, any one of the companies' main operating cities – Rome, Paris, New York or Sydney and working on programme material to be transmitted in a dozen different countries.

The closed days of the BBC and ITV duopoly have gone forever. Even foreign takeovers of ITV stations have been discussed at the highest level. Sir George Russell, the Head of ITC, warns that when the moratorium on takeovers in the UK ends in 1994, the huge combines in Europe – Bertlesmann of Germany, Hachette and Havas in France, Polygram in Holland and Fininvest in Italy – are each larger than the entire advertising revenue of the present 16 ITV companies.

The larger ITV companies are already lobbying the government to relax the rule about takeovers to enable them to absorb the medium-sized companies. This might make later overseas infiltration more difficult. □

The latest companies to be granted franchises by the ITC have adopted very restrained on-screen idents. GMTV's version is redolent of the world of the advertising agency rather than broadcasting

Chapter 3   Patterns of Graphic Skill

What do graphic designers working in television need to know? What skills have they to acquire in their formal education and in their professional lives?

Having employed hundreds of graphic designers, over a period of more than 20 years, Thames Television decided in 1989 to produce an internal company report to find answers to these seemingly naive and open-ended questions. The final paper was daunting in its range and detail.

Reading the ensuing 14-page list of the skills and knowledge thought essential to the work of the tele-vision graphic designer created a vision of a cross between Frankenstein's monster and the Admirable Crichton. Attempts to formalise categories of human performance often fall into this mode.

Graphic expectations

Expectations encompassed in the Thames study were as follows. Under the heading 'Skills' were 'hand/eye/brain co-ordination' and 'adaptability/flexibility'; under the heading 'Knowledge' were 'the laws of copyright and libel', 'union agreements' and 'company admin-istrative agreements'. (It is easy to imagine the sub-versive documents that might be circulated by employ-ees after reading such material.)

The report was a much needed exercise to codify the whole area of graphic design work at a time when the range of responsibilities appeared to be endlessly complex – sprawling out of control.

Here is a précis of that work's widely applicable descriptions. These notes were originally intended to construct a programme of internal training. The envi-ronment and working framework at Thames Television at the time would have been much the same in other graphic design departments within television.

The skills required

All graphic design staff or applicants for interview were assumed to have completed an art foundation course and three or four years' training at an art college, almost certainly in graphic design.

This would have given them the fundamental basis of *design* ability. Some graphic designers-to-be may have had a passion or obsession for television design, film-making, or animation, from their early teenage years.

The growth of higher education in art and design over the past 30 years has made the chance of progress for an outsider, maverick, or 'talented tea-boy' less and less probable.

The first heading in the Thames report was 'The operation of equipment'. All graphic designers were expected to be able to operate *all* the graphic design equipment in use in the company 'correctly, accurately, speedily and cost effectively'. At that time the graphic department housed: (a) a *camera system* for photograph-ing artwork on to 35 mm slides where pin-registration mounts would allow perfect on-screen alignment and transfer through tele-cine slide scanners within the safe transmission area; (b) a unit for making *dry-transfer* (rub-down) lettering and symbols; (c) *35 mm stills cameras* – for reference work and for making story-boards; (d) a computer-based *digital paint system* (the version in operation at that time was Quantel's 'Paint-box' and all graphic designers were given a formal training course on the machine at Middlesex Poly-technic); (e) an *electronic video rostrum camera*, attached to the 'Paintbox', to enable them to input artwork, photographs, colour transparencies and three-dimensional objects; (d) any *character generators* instal-led in the company studios; (f) all the company *video tape recording systems* used in parallel with graphic design production (Umatic, VHS, BETACAM, and the then current in-house standard M2); (g) the *electronic library system* used in-house which supplied stills, from a wide range of sources, within the electronic graphics workshop at their London studios.

From this section alone it will be clear that the boundaries of the work in graphic design were being extended well beyond film animation; they were becoming more deeply interwoven with television's complex electronic routines for transmission.

The major questions in the air were: how much was the graphic designer's role to change – from artist to operator?; Were they to be art directors or technicians?; and will the new technology improve or diminish design standards? These were unlikely to be resolved quickly or definitively.

Patterns of knowledge

Another section in the Thames report listed 'the areas of knowledge' that graphic designers in television were required to absorb. They were: (a) the *capabilities of all the equipment* mentioned above and an understanding of what they are best suited to achieve; (b) a clear knowledge of the *technical and engineering production chain* relating to the transmission of graphic images and animation; (c) the uses and capabilities of the *graphic effects machines* (e.g. 'Charisma', 'Ampex ADO', and 'Encore', etc.); (d) the application of *two- and three-dimensional computer graphic systems* available in-house and at facility houses; (e) *desk-top publishing systems*; (f) the operation and potential of *video and film rostrum cameras*; (g) all the *studio and gallery facilities*.

Under the sub-heading 'Photography and filming' graphic designers were expected to be able to apply these techniques to their daily work and to: (a) assess where photographs are technically suitable for transmission; (b) choose the most appropriate film or video equipment for the design in production; (c) prepare and use mattes and masters for film artwork; (d) apply registration and the use of field-sizes to rostrum camera work and be able to carry out hand-drawn animation (i.e. prepare keyline drawings, paint and trace cels, rotoscope from film or video sources, prepare film material for back-projection or aerial-image, slit-scan opticals, and make competent dope sheets for rostrum camera operators); (e) deal with photographic and film processing; (f) know the capabilities of high-speed and time-lapse and other cinematographic effects (wide-angle, snorkel and endoscopic lenses, etc); (g) understand the potential of various film speeds, paper stocks, filters and effects; and (h) know what specialised facilities there are for film and video production.

A section was devoted to the graphic designer's relationship to *model-making* (sometimes it's DIY!) and to the filming and video animation of models – from stop-frame to computer-controlled motion rigs. Ensur-

ing models were made to the correct degree of detail and finish for filming was stressed. For some stop-frame model animation the design of controllable armatures was an important aspect.

Competence was also essential in an unexpected area – *design for print*. Television graphic designers have to apply all printing techniques (lithography, silk-screen and letterpress, etc). They need to prepare instructions for printers and typesetters and to proof-read satisfactorily. All this is for the production of 'set-props' used in dramas, situation comedies and documentaries. At some time every possible kind of printed item has been called for: posters, magazines, newspapers, packs, menus, tickets, and print ephemera of every kind – in any language – for any era. Graphic designers are also responsible for the design and *production of set-dressings* of this nature. Great care, research and observation are expected when making such items as genuine as possible. Research and knowledge of contemporary and period lettering are equally important in the design and specification of signs when working on locations with set designers and location managers. Designing the house-style for a 1930s shop in Manchester or a menu for an expensive New York restaurant in 1960 means you must know your Garamond from your graffiti.

A great deal of graphic design is applied to *studio settings*. For this it is essential to work with the set designer, to be able to read and interpret scripts, then to create or enhance the scenery through conventional construction and painting, or to use film or electronic graphic effects for all types of studio-based programmes. Documentary, quiz, discussion and feature programmes are the kind most often to involve graphic design in this manner.

Music and movement

No graphic designer working in television will cope very well without a deep concern for the soundtrack. Choosing and planning the choreography of multiple images and then briefing a composer to write original music to fit the storyboard, or to being able to manipulate images to synchronise with given music, are both essential formulas. Ideally the designer is expected to read music! But a knowledge of music libraries, copyright, agreements with the Musicians Union, and as wide as possible an interest in all kinds of musical instruments, titles of music, singers, performers and composers were all an expected part of the designer's repertoire according to the Thames study.

Team-work

The last section in this description of the paragon among graphic designers for the small screen explains the importance of *working effectively with all the other members of the production team*. Beginning with the relationship to the producer and/or director the roll-call here involved writers, film and video editors, camera operators, lighting directors, make-up artists, costume designers, and the operators of all the electronic effects equipment available in-house and externally.

Organisational sense, in order to co-ordinate each project, is vital and the *control of the budgetary system* (which will vary from one company or institution to another) will always be a major asset. Effectiveness in negotiating the purchase of services is increasingly important. Individuals vary of course but there is a hope that graphic designers will be valued for their creativity, rather than simply for 'coming in just under budget'!

Summary

Here is a résumé of those qualifying skills and areas of knowledge, idealised or not.

1 *Art education* to BTEC Diploma or BA (Hons) level, almost certainly in graphic design.
2 *Operator skills required*
  a A camera copying system
  b Dry transfer production unit (rub-down lettering)
  c 35 mm stills cameras
  d All digital paint systems held by the company
  e A video rostrum camera
  f Various video tape recorders
  g An electronic stills library
3 *Areas of knowledge*
  a To know the capabilities of all equipment
  b All aspects of technical operations (affectionately known as 'Tech Ops') and engineering routines for graphic design production and transmission
  c The use and capabilities of graphic effects devices inside or outside the company
  d The application of 2D and 3D computer graphic systems used in-house and in facility companies
  e The use of desk-top publishing systems
  f The routines for preparing artwork for video and film rostrum camera sequences
  g The use of studio, gallery and technical facilities
4 *Other areas*
  a Good inter-personal skills
  b An appreciation of time management
  c An awareness of client care
  d Adequate negotiating skills
  e Construction and animation of models
  f Typography and design for print
  g Relationship to set design, scenery construction and the set designers
  h Application of music to animation and pro- gramme-making
  i Ability to work with, and a thorough knowledge of, all other grades and skills used by internal colleagues and outside services
  j The responsibility to manage graphics budgets
  k Comprehensive knowledge of stills, photographic and film processing

Manager, accountant or artist?

When this inventory is reviewed there is a feeling that the real qualities may have been omitted. These are the gifts of the 'three Is': intuition, inventiveness and imagination, plus the desire to amuse, instruct and entertain – elements which make or break designers and artists in all fields. These are the qualities most difficult to evaluate and teach. Walter Gropius, Principal of the Bauhaus from 1919–1928, said 'Art cannot be taught'.

The paramount skills

The most often repeated attribute expected by those inside the profession, of both their fellow designers and those wishing to join it, is the *ability to draw*. Any definition of what this means in terms of style or quality will be disputed and confused, but the phrase 'a fluency to depict ideas' might just resolve a long debate. In a feature on high-technology and art education in the *New Statesman & Society* Beverley Russell quotes Colin Forbes, a founding member of Pentagram, as follows, 'Learning to see was my most valuable experience as a student. Drawing is not a skill with the hand; the reason you can't draw is because you can't see. Today I still draw out ideas when I'm working.'

At a careers advice session presented by the Chartered Society of Designers some years ago the panel members, all designers of some kind, were each asked what they thought was the most esteemed asset for any designer. Sir Hugh Casson said 'Genuinely to like people'. Unexpected, but wonderfully comprehensive.

Ways in which those wishing to work in graphic design for television might gain some of this knowledge and experience through art education, and later by training in the industry, are outlined in Chapter 5. □

Chapter 4   Graphic Interpretations

If graphic designers working in television were ever to get an inflated idea of the value of their work as communicators they might soon be brought to ground level by comments like the following newspaper review of a BBC2 *Horizon* programme by Thomas Sutcliffe: 'This was a fascinating programme (only mildly obscured by its explanatory graphics) . . .' (*The Independent* 7 April 1992).

It does suggest that graphic design should avoid being irritatingly conspicuous.

There is a clearly detectable movement away from fads and effects. A return to ideas and content is called for. In his third book, soon to be published by Yale University Press, *Design, Form and Chaos*, Paul Rand scathingly describes much graphic design of today as:

> this bevy of depressing images . . . a collage of confusion and chaos, swaying between hi-tech and low art, wrapped in a cloak of arrogance: squiggles, pixels, doodles, dingbats, ziggurats . . . Art Deco rip-offs . . . convenient stand-ins for real ideas and genuine skills.

His accusation, broadly aimed at typography and design for print – the area he has practised for so long – applies to much of the torrent of graphic design produced for television. How would Rand assess the design we see on our screens? It would be interesting to hear. His desire for content should not go unheeded.

Opening titles

The three main areas of activity – designing opening titles, contributing visuals to programme content, and design for on-screen promotions – have remained unchanged since television transmission began.

'Programme main title design is where the graphic designer's talent and skill are at their most visible.' So says the cover of a BBC Television's Department of Graphic Design showreel.

Making *programme titles* still captures the attention and dominates the awards scene, so, to the enquiry 'What do graphic designers actually do?', the clearest reply is to refer to opening sequences.

Fewer drama series and large-scale documentaries have been made over the past five years, or so. These usually provided long-term projects with carefully planned funds, giving time and reasonable resources to creative and technical staff. Recently resourcefulness in stretching low budgets to achieve the almost impossible is the common story behind much of the work seen on even the most distinguished showreels. Design is about using resources well, so perhaps all is not lost.

Promotion work

Clear brand identity is now crucial in the competition between channels in the UK. BBC1 and 2, the ITV network, and British Sky Broadcasting are all extremely aware that graphic design helps to sell programmes to their audiences. All channels have engaged graphic designers, in-house and from the independent groups, to design and produce the ever-increasing volume of *station idents and on-screen promotion* work. The intensifying of promotional effort led Ian St. John, then a manager at Lambie-Nairn & Company, to say in an interview with *Creative Review*: 'This trend will make British television even more polished than that in other parts of the world.'

In on-screen promotion work there are short sequence 'stings', the minimal 'mnemonics' and longer and more elaborate future programme information 'rundowns'. All these are using graphic design more effectively. BBC2's *nine* variations of an animated ident is a clear example (an account of the design and production of these appears in Chapter 7) and additional in-house versions have now extended the themes. These are very different from the repetitive, sacrosanct versions of the past.

The term 'mnemonic' is a comparatively new one in television design. It means a device to aid the memory.

As the earliest Greek and many later forms of mnemonics rely on *visual* aids, it is an appropriate name for the two- or three-second animations which emphasise tradenames or symbols.

When English Markell Pockett produced a series of branding animations to be used by all the ITV regions in 1990, subjugating their separate station idents, Richard Markell was quoted as saying 'Like any packaging, we're putting ITV on a supermarket shelf along with other channels.'

Programme graphics

Design for *programme content*, a third category of graphic work, has also increased over the years. Faster and efficient means of delivering images has found programme-makers asking for further graphic input and being willing to involve graphic designers in the visual styling of *whole* programmes. Information graphics in documentaries, features, and news-based programmes on every possible subject have added a higher profile to graphic activity. Designers have adapted to digital paint systems, character generators, new editing equipment, and the management and daily operation of stills libraries. The output is vast.

Storyboard to screen

Procedures in design remain constant in spite of equipment changes. Briefing can vary from a hurried telephone call, made in a state of semi-panic, to a day-long meeting with a 50-page survey of the presumed audience profile and the technical operations. Graphic design for the Olympic Games or an election involves months, even years, of team-work.

Poor instructions are an everyday hazard. After waiting for information for many weeks one graphic designer became rather cynical. On being challenged as to why he had not started, he replied, 'Until the programme people tell me what they want I can't possibly tell them they can't have it!' Briefs saying, 'Do something exciting like the BBC News and *The South Bank Show'*, or 'We would like to use that wonderful Japanese computer effect they have on *Tomorrow's World* ', are not much help.

Just as an item of industrial design would not be made without sketches, colour samples and measured drawings, or a house without plans, the storyboard is the essential starting point. These are now likely to be prepared with digital paint systems. Off-screen colour photographs (taken with a Sony or Hitachi camera unit) are then pasted on to a board. Details of the soundtrack and other information on the movement and intended effects, to make the board as 'graphic' as possible, can then be added.

Producers and directors have learnt to react well to these static and mute versions of the designer's intentions. For prestigious work animatics can be made.

These are recorded sequences of stills, most likely shot on a video rostrum, simulating animation with the sound, if possible. Commercials and broadcast titles both use this method.

Approved storyboards have other functions; they present the basic information for those involved – model-makers, film camera operators and computer programmers – to assess the most suitable production methods and are the basis for estimating the production costs.

Graphic designers' aims

What do graphic designers see as their main objective? Working closely with a director or producer as their client, they must first pay attention to the overall concept of the programme and translate that into visual terms. They must absorb themselves as much as possible in every aspect of the programme's background and be on fluent terms with the editorial team who usually have the advantage of having worked on the project for a longer time.

Illusion in some form, as stated in the introduction to this book, is the core of all pictorial presentation. Artists and designers strive to re-create the real world, or go beyond 'reality', employing devices to engage the audience. This is not alluding to optical illusions but to the whole basis of graphic design. Professor Martin Kemp's book *The Science of Art* (Yale University Press, 1990) presents a wealth of images on perspective and all forms of image-making. He records the attempts of engravers and photographers, and artists of great stature, from Brunelleschi (who is amusingly described by Professor Kemp as 'unusually well educated for a practitioner in the visual arts') to Seurat, to pursue scientific methods to deceive the eye. The geometric studies of perspective projection by Albrecht Dürer, and others, made around 1500, established methods now fundamental to ray tracing. These were also remarked upon by Georg Rainer Hofmann of the Fraunhofer Computer Graphics Research Group in an article 'Who

al scaling. Putting this as a central theme indicates the need to move from readily available subscales to the general or global level. The focus is on the unresolved and often overlooked issue of how to deal with the extracted factors or components, regardless of how well a latent structure could be constructed and confirmed. The primary purpose of the Chen Approaches to Unidimensionalized Scaling (CAUS) is to provide some guidelines for the advancement of scientific standards to promote and direct exploration in scaling practice. The following is a brief description of the CAUS in terms of a theoretical and a practical approach.

Theoretical approach to unidimensionalization

Chen's theoretical approach to unidimensionalization is formed by learning from science, specifically from engineering mechanics. You may find it interesting how theories of mechanical engineering would contribute to our knowledge of scale development in social and behavioral research. Yet, not only was the history of psychometrics in large part a history of psychophysics but a great share of the major ideas in behavioral and social research came from various pure and applied fields of science. Specifically, concepts such as "stress," "support," "functioning," and "mechanism" all find their roots in engineering mechanics. However, when these terms were grafted to psychosocial studies, their theoretical underpinnings were not all taken along with them. It is understandable because it is not easy for behavioral and social scientists to grasp all the fundamentals of fields other than their own. However, it may have left substantial gaps in knowledge. Here, a review of measurement-related theories in engineering mechanics will provide some useful insights.

Engineering mechanics, especially its basic component material mechanics, can be regarded as a science of stress of structural parts connected together in such a way as to perform a useful function and to withstand externally applied loads. The kinds of load include press, pull, and shear, which subject different parts and structures of various materials to tension, compression, bending, and torsion. This results in a balance represented by various external forces that cause the strain and stress in the material. Stress is defined as the intensity of internally distributed force, which, marked by the strain in the material, is a reaction to the external force. The study of strength of materials is aimed at predicting how the geometric and physical properties of a structural part will influence its behavior under service conditions (Timoshenko & Young, 1962). Whether the structure is strong enough to withstand the loads and stiff enough to avoid excessive

invented ray tracing?' in the journal *The Visual Computer* (June 1990).

We are all intrigued and deluded at many levels by graphic presentation. A few of the endless ways in which this is effected in animation are revealed in the 'case histories' in Chapter 7.

Television 'typography'

How did it grow? Set and costume designers merely continued their roles from the cinema and theatre – but a 'graphic designer'? No one was really sure what might be required. So it was from the truly 'prosaic' need for on-screen lettering that the profession arose. Graphic design will always be concerned with the presentation of words.

Designing with type in television was the subject of a lecture given by Alan Jeapes at a conference presented by The Monotype Corporation in 1990. Jeapes is one of the most experienced graphic designers at the BBC. In attempting to identify two 'styles' of television lettering he compared them to those of the tabloid press and the Sunday supplements:

> I think that the designers responsible for fine and exciting new wave typography in the world of print are not the same people responsible for the 'folk' end of the business. They are different animals. However, in television they are often one and the same. A designer working in television may be commissioned to work on a serious classical piece one day and a game show the next. They may develop what I can only describe as 'typographic schizophrenia'.

Television typography, Jeapes emphasised, is 'about headline setting . . . it can never approach the joys and subtleties of body matter in print'. The tiny amounts of copy to be presented on-screen at any one time is a major shock to designers from the world of print. The few lines of type that fill the screen and the time required to read the words are very limiting factors. With little copy to display and the constant screen proportion of 3:4, there is less scope for the nuances of size, weight of face, and style so precious to the typographer of the book, magazine or newspaper. This is not to propose that immense care is not taken with lettering in television. In information graphics for news, documentaries and elections a very high level of typographic presentation is achieved now that good resolution and anti-aliasing are standard.

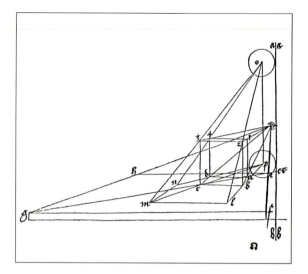

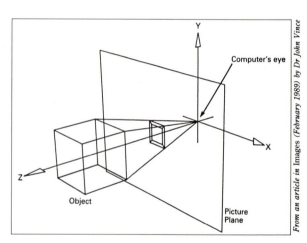

From an article in Images *(February 1989) by Dr John Vince*

The laws of light and perspective, investigated 500 years ago by Dürer and others at that time, still apply to computer animation in the fascinating area where art and mathematics can be seen to combine usefully in a timeless manner

Even for the computer, plotting perspective is managed via an imaginary window, using x, y and z co-ordinates

One commentator, the American-born typographic historian Beatrice Warde, hoped that designers 'will not be too much affected by attempts to imitate the electric mobility of letters on the screen'. Graphic designers in television might be reminded to restrain their animation of lettering.

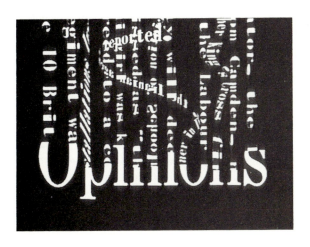

Franco Grignani's cover for Typographica 16 *demonstrates one of his many designs where type appeared to flow across the printed surface in the 1950s and (below) a still from Bob English's title for* Opinions *with dynmnic lettering that won him an RTS Graphic Design Award nearly forty years later*

Animating type in titles

The title of the programme is the essential element of every opening sequence and its treatment can be the crucial point in success or failure of the design.

Analysing just over 30 titles on a BBC showreel revealed that the mean time for the on-screen appearance of the title lettering was only one-tenth of the whole animation – just two to three seconds. The origination of the letterforms was wonderfully diverse and very difficult to discern by observation alone, but was identified as dividing into four methods: computer-generated and animated, formed as a model, hand-drawn, or set by conventional print-based systems.

Two titles where lettering is animated in a restrained and purposeful way come from the same source – English & Pockett. *Screen Two*, designed by Darrell Pockett, opened with the subjects dividing in two as mirror images. These split-screens were created in post-production on 'Harry'. Each letter of the word 'TWO' then grew, aptly and elegantly, from two separate elements. Bob English made *Opinions* for Clark Productions. It was shown on Channel 4 and gained a Royal Television Society Graphic Design Award and a Silver D&AD Award in 1989. The dynamic typography could not have been achieved without electronic technology. An inability to draw moving lettering in cel animation was a great constraint. Franco Grignani, an Italian designer/typographer, used twisting letterforms made by distorting photographic negatives in his two-dimensional graphics in the 1950s, and his cover for the magazine *Typographica 16* has a strong resemblance to this later work. Influences and inspirations, conscious and unconscious, are explored in Chapter 9.

Although there are occasions for more exuberant games with letterforms, these two examples of gentle and almost unobtrusive animations aspire to the work of the best typographic designers of the past.

In a passage of plain speaking, Alan Jeapes also said: 'Those of us who work in television know what an appalling sausage machine the business is and that the graphic designer's efforts can be easily ignored' He reflected on another aspect of designing for what he called 'the momentary existence of the televised image'. Did it breed, he asked, 'an uncaring cynicism in many designers?' It can, but it is reassuring that these thoughts came from a designer whose work has, for 30 years, searched for refinement of typographic detail with enthusiasm.

Working in television is certainly very demanding and only the most dedicated will persist and succeed. □

Chapter 5   Training for Television Graphics

Training for all professions divides into two streams – the stream of formal education, which comes to an end, and the stream of learning within each profession which necessarily should continue throughout everyone's career.

Education and television graphics

From foundation courses to specialised studies in every branch of art and design, then on to MA degrees in highly specialised sections of these subjects, the art schools, colleges, polytechnics and now universities in this country *can* provide the environment and stimulus to train for all branches of the design profession to a very high standard. Given the resources, they have done this as well as anywhere else in the world. The increasing influx of students from America and most other countries testifies to this.

There have been advances and much re-structuring in design education during the past quarter of a century (not all of it productive), in adapting to movements in various branches of the design world.

Few colleges have been transformed during this period to the extent of the extraordinary advances that have occurred in the telecommunications and entertainment industries.

How the need for designers grew

Opportunities for prospective graphic designers have increased immensely as television transmission and production output has grown from the early days of the monopoly of the single-channel BBC. This was followed by the dual-channel environment of the ITV years, extended later by the start of BBC2. Channel 4 brought further opportunities for graphic design and, as stated, its unique way of operating fostered many of the independent production companies, who in turn employed graphic designers as freelance contributors.

Another expansion was the setting up of the Astra satellite stations. These started with two contenders – British Satellite Broadcasting and Sky channels – now reduced to British Sky Broadcasting. There is the anticipated launch of Channel 5 which could create fresh opportunities.

Until 1955, with the sole output from the one BBC channel in this country, there was little evidence of a clear vision to recognise that design (not solely graphic design) would become an essential factor. New visual territories to explore would need new training.

No special training

For the handful of art school trained designers who first entered the BBC to form a properly constructed Graphic Design Department, and those who joined the newly created ITV companies (ATV, Associated Rediffusion and Granada among them), there was almost no awareness of the use of graphic design in television or any study of the subject. Hardly any students of that era had ever considered their career might be in television.

They arrived in the studios completely uninitiated, although some may have shown a personal interest in or even experimented with film-making. Less certainly, a few may just have been introduced to hand-drawn cel animation and been aware, from the cinema of their childhood (perhaps through the cartoon films of Walt Disney), that you could make drawings move and that *somewhere* artists were employed in a magical and glamorous world called the 'film industry'. The new recruits to the infant industry of 'TV' had no specific training or knowledge of the specialised activity they were destined to operate. 'Storyboard', 'rostrum camera', 'field-sizes', 'pic-sync', 'movieola' – the vocabulary was a secret code.

Their undoubted success was due to their thorough grounding as artists and illustrators; as well as their competence to adapt, they could *draw*. The lesson is that the fundamental purpose of an art college education is to inspire and develop students as *broadly as possible*, and that too much over-specialisation is unwise.

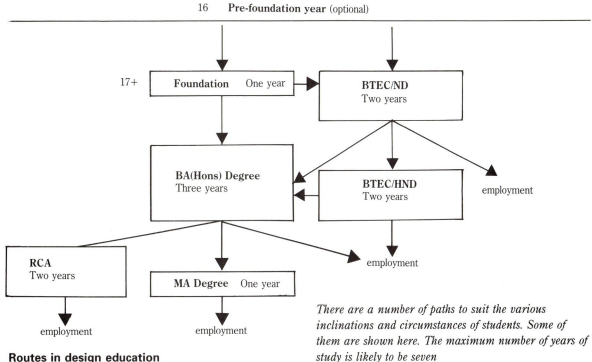

16   **Pre-foundation year** (optional)

17+   **Foundation**   One year   →   **BTEC/ND**
Two years

BA(Hons) Degree
Three years

BTEC/HND
Two years   employment

RCA
Two years   **MA Degree**   One year   employment

employment   employment

Routes in design education

There are a number of paths to suit the various inclinations and circumstances of students. Some of them are shown here. The maximum number of years of study is likely to be seven

The constant acceleration in the changes in technology, equipment and the realignment of the very core of the broadcasting in this country reinforce this assumption.

A *New Statesman* education feature (quoted in Chapter 3) found that leading British designers, many of whom had received their own art education in the 1950s:

> ... soundly dismissed a narrow focus on vocational training at a time when the world is changing so rapidly that what is taught as relevant one year may be obsolete five years down the road. The curriculum should be broad-based and point towards flexibility in the student's career.

Too much specialisation?

On stating that over-specialisation is unwise, it might seem contradictory to guide any graphic design student to concentrate on graphic design for *television*.

They will find great difficulty in discovering courses to suit them. The general printed information on the courses in the various colleges conceal, rather than

reveal, an acceptable route. The full prospectus from any college will give reasonable details but it is not easy to pinpoint references to graphic design for television. Obviously, lecturers do give valuable advice based on their personal experience, but it must be remembered that television is a minority option, and often their own knowledge of this area may be limited.

In the foundation 'year' students are expected to visit as many prospective college 'open-days' as possible to ascertain the course content, staffing and facilities for their BTEC or BA course. These vital visits usually take place in the month of February.

The essential route

The chances of being interviewed for even the most junior design post at any television station, or by any of the now well-established design groups specialising in television graphics, without completing an art school course would seem almost impossible. The only exception might be someone who has learnt to operate one of the electronic devices – a paint system, a digital effects machine, or a character generator. (It is interesting to note that some recent on-screen credits have said 'Harry/Paintbox *Artist*'.)

Even then, their employment and status as a potential *designer*, rather than as an operator, would be very doubtful. More and more design-trained and fully experienced television designers have embraced, or have been coerced into, the arms of the electronic apparatus. There is no implication that designers are of more worth or of higher status than operators – they have simply opted to do a different job. The experience of working with imaginative technicians is an immense pleasure.

The diagram on the previous page shows the various pathways at present on offer in art education. As increasing numbers of students from abroad come to study here, at all levels, from vocational courses to BA, MA and PhD, it is important to repeat that the competition to enrol on already very crowded courses is very high indeed. ADAR say 'Your wish to be considered for a Diploma or Higher National Diploma place is shared by many thousands of young people. It is important therefore that you select your choices wisely.'

Are there firm foundations?

Foundation courses in art schools, under the present system, are demonstrably very short.

They are almost invariably for only one year. (There are some two-year courses but these are part-time and intended for mature students.) Arriving in September for the new academic year, students are, within a few weeks, even before the end of their first term, expected to decide on their area of further study for the remaining two terms – fine art, graphic design, fashion, textiles, furniture, interior design, ceramics. Many do not even comprehend what the options might be. Early in the new year they must make application to one of the institutions where they hope to follow the discipline they have opted to study so abruptly. 'Discipline' and 'institution' – both have such awful overtones – but they are the words in present currency.

Compressed into the bare 10 weeks of the first term of foundation studies are life-drawing, studies in colour theory, perspective, the history of art and design, general studies, general drawing and painting, theories of design, composition, ratio and proportion. A background of information about the potential areas of professional opportunities are poured on top of this concentrated diet. Sometimes lectures by practising artists, photographers, illustrators and designers in particular areas may be given to clarify at least some of the opportunities for careers in advertising, magazine design, publishing or even graphic design *in television*!

In graphic design courses the options available, as set out in the book published each year by the Design Council entitled *Courses in Art and Design*, are profuse. Students seem to decide on the direction of their future work with minimal time to evaluate their own strengths and interests or to be able to discern the variety of possible professional destinations. The *end* of a foundation year would be a more reasonable stage at which to commit themselves, but processing many thousands of applicants in weeks rather than months for the next academic year makes that untenable. So why not two years? In the past there had been three years of basic study. People then had a far better chance of resolving which of the many design courses they were best suited to pursue.

Anyone who can describe the contents, facilities, staffing or distinct attributes of more than a handful of design courses would be rare indeed. These Cassandra-like thoughts are to underline the complexity of choice that has to be faced. They are primarily to emphasise that the difficulties of this early stage are made much worse by the need to make commitments in weeks. Ironically the *PCAS Guide for Applicants* (Polytechnics Central Admissions System) says in the preface, 'Considering a course of higher education is a major step forward. *It's certainly not to be rushed*.' (My italics).

Vocational training

Even the best career advice at secondary education level faces a daunting task in merely setting out what opportunities exist. Once the decision is made to study art there is a huge choice but many applicants arrive for interview with 'a very confused idea of the nature of the design disciplines. Most however have heard of graphic design and assume that this is the only design subject' (Farnham College, 1990).

Entry qualifications

The educational system offers what might appear to be a direct route. For those who feel a less academic path suits them there are BTEC NDD and HND courses. These are most frequently commenced without a foundation course when students are a minimum of 16 years old and have four GCSE passes at C grade or above. Some colleges will consider less. In a report on their BTEC

NDD course in 1990 West Surrey College of Art and Design at Farnham said 'One or two entrants have shown exceptional ability and . . . the Course is able to "rescue" some students for whom the traditional educational pathways have proved inadequate.'

'Late starters' and 'mature students' are able to join design courses. Many have done so with spectacular success. Some may already have high educational qualifications; others will have to work to gain them.

There seems to be a trend towards a 50 per cent proportion of female to male students – removing the male majorities of the past.

The routes beyond the two years BTEC at National Diploma level divide into either direct employment or further educational courses. As many as 80 per cent will opt to continue their full-time education.

The Business and Technology Education Council validates courses in graphic design at many schools, fairly evenly spread throughout the country. A list can be obtained from them (see address index, pp. 139–140). They award the BTEC National Diploma on the successful conclusion of a two-year full-time or a three-year part-time/sandwich course. Students can use this as a basis to apply for entry to a first degree course (in place of a foundation course), or they can choose to study for the BTEC National Diploma in Design, two years full-time or a BTEC Higher National Diploma which demands two years full-time or three years part-time/ sandwich studies. Sandwich courses include time allotted to professional practice attachments.

Choosing a course

There are around 60 colleges running these general graphic design courses. Their content and amount of specialisation varies greatly. The publication *Design Courses in Britain* (mentioned above) gives the good advice:

> General design courses usually include a wide range of subjects for practice and study related to graphic design. Most enable students to give equal attention to a variety of specialist disciplines, but it is often possible to specialise after the first or second year in one or more aspects.

Finding a college course that concentrates on *graphic design for television* is not simple. In 1992, the index in the annual ADAR guide to higher level courses in art and design does not even list the subject!

'Animation', 'Computer Graphics', 'Computer Image', 'Time-based Media' and 'Film and Television' are given and these may be a route into television for some. At present only Ravensbourne College of Design and Communication provides professional level electronic equipment within a specific course in graphic design for television at BA (Hons) level.

Some colleges provide support for students interested in television graphics because their staff have had past or present involvement in television. The BTEC course at Suffolk College, in Ipswich, has this tradition and Alisdair McMurdo, an ex-television designer and his colleagues have stimulated students interested in television in the past few years with great success. Kingston University under the guidance of an ex BBC graphic designer, Jean Braid, encouraged students in this field over a long period. Many of them are now well established in the industry.

The degree route

Degree courses were, until 1992, validated by the CNAA (Council for National Academic Awards). New arrangements are now being made whereby this institution is to be disbanded. As some colleges of design are now part of polytechnics (many recently given university status) or institutes of higher education, their courses will be self-validated and degrees awarded by the governing body. The annual ADAR publication gives a full list of degree and HND courses in design. The address is given at the back of this book.

General graphic design courses will enable and encourage students who develop strong interests in a particular area to follow them as much as possible. The choice of subject for their dissertation in cultural studies and certainly the major project in the final year should, and do, come from student motivation.

The BA (Hons) degree will take a minimum of three years full-time or four years for a sandwich course. Beyond this first degree level there is the opportunity for an increasing number of students to obtain an MA in one or two years at selected colleges, or for them to gain entry to the Royal College of Art where a Master of Design (MDesRCA) can be conferred.

Facilities, equipment, staff and course structures can never be totally stable. Those which stand out as places where students have been able to acquire a *strong interest in graphic design for television*, as well as ADAR listed courses in animation and computer graphics are shown overleaf:

Graphic Design Courses in the UK

Some colleges listed here do specify 'TV Graphics' and the 1993/4 ADAR guide points out that modular courses are presented by more and more colleges, enabling students to choose a path 'nearly tailor made for your own particular strengths and interests.'

A = Animation course listed by ADAR
CG = Computer graphics course listed by ADAR
VA = Video animation equipment

			Anglia Polytechnic
	CG		Barnet College (with Middlesex University)
		VA	Bath College of Art and Design
			Berkshire College of Art and Design
A		VA	Bournemouth and Poole College of Art and Design
		VA	University of Central England in Birmingham (was City of Birmingham Polytechnic)
		VA	University of the West of England at Bristol (was Bristol Polytechnic)
		VA	Camberwell College of Art
		VA	Cambridgeshire College of Arts and Technology
		VA	Carmarthen College of Technology and Art
		VA	Cumbria College of Art and Design
			Central Manchester College
		VA	Central Saint Martins School of Art and Design
	CG		Cheltenham and Gloucester College of Higher Education
		VA	Cleveland College of Art and Design
		VA	Clwyd College of Art and Design
	CG		Colchester Institute
		VA	Cornwall College/Redruth
		VA	Coventry University (was Coventry Polytechnic)
		VA	De Monfort University (was Leicester Polytechnic)
		VA	Duncan of Jordanstone
		VA	University of East London (was Polytechnic of East London)
		VA	Epsom School of Art and Design
			Falmouth School of Art and Design
			Farnborough College of Technology
		VA	Glasgow School of Art
A		VA	Gwent College of Higher Education

A		VA	Harrow College of Design and Media at the University of Westminster (was Polytechnic of Central London)
A			University of Humberside (was Humberside Polytechnic)
			Kent Institute of Art and Design/Maidstone
		VA	Kingston University (was Kingston Polytechnic)
		VA	Leeds Polytechnic
		VA	London College of Printing
			Manchester Polytechnic (new University status)
A		VA	Middlesex University – MA in Computer Graphics – (was Middlesex Polytechnic)
	CG		Nene College
	CG		Newcastle College
A	CG	VA	Northbrook College of Design and Technology
A			University of Northumbria at Newcastle (Newcastle-upon-Tyne Polytechnic)
		VA	North Staffs Polytechnic
			Norwich School of Art
	CG	VA	Ravensbourne College of Design & Communication
		VA	Royal College of Art
A		VA	Suffolk College of Art and Design
		VA	Walsall College of Art
A			West Sussex College of Art
		VA	Wolverhampton University School of Art & Design (was Wolverhampton Polytechnic)

Most courses, both BTEC and BA, give students some introduction to graphic design for television and they will encourage them to pursue the subject if it becomes their major option. Colleges with film, video or computer animation facilities are shown

Animation as a starting point

A strong instinct for animation and the moving image has led many design students towards television graphics. They may not develop the high skills of the best hand-animators, but experimenting in all manner of ways at an early stage is very rewarding and many colleges have equipment that fosters animation. Video has provided a 'hands-on' climate that was not so easily achieved in film.

Equipment and the cost factor

High capital costs of television equipment are generally beyond the budgets available in education. A single character generator will cost around £15K. The industry standard video editing systems are in the region of £200K. A digital paint system of reasonable scope can be purchased for as little as £5K but the type used widely in television production are about £80K plus stills stores and 'extras'.

Low-cost start-up

Electronic equipment is provided in many colleges of art and design., at a modest level, to give students an introduction in to what is termed 'time-based media'. Harrow College of Design and Media (now part of the University of Westminster) has, for example, a unit containing two video rostrums with EOS900 animation systems, one high- and one low-band Umatic. The EOS900 is a broadcast standard colour animation control system used widely by professional animators and a description is given in Chapter 6. The EOS Animation Controller 580 is a less expensive system. (Colleges with similar animation facilties have been indicated in the list above with the initials 'VA'.

For inexpensive paint systems Harrow College has four Commodore Amiga computers, at about £1.5K each, using frame-grabbing through a Genlock which allows picture signals from the 'home-computers' to be recorded with sequences from live-action and other video sources. There are also two Apple Macintosh units. All of these provide some animation facilities.

The senior lecturer, Al Morrison, considers these are an ideal starting point, as they are easy to follow and the Commodore is relatively fast in response.

The Harrow animation facilities described here have been used by students at all levels in the past few years, and those at advanced stage have gained the Benson and Hedges Student Gold Medal in Video – Susan Hewitt in 1988 and Samantha Padget in 1989.

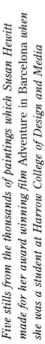

Five stills from the thousands of paintings which Susan Hewitt made for her award winning film Adventure in Barcelona when she was a student at Harrow College of Design and Media

When Susan was on the illustration course at Harrow College she produced her eight-minute hand-drawn animation entitled *Adventure in Barcelona*. This was a sequence following a day in the life of the famous street, Las Ramblas. For this she made around 6,500 separate watercolour paintings. These were recorded double-frame to a music track on one of the college's two video rostrum cameras using an EOS animation controller. Her reward was the first prize, £1,000, in the Benson and Hedges Animation Competition.

Another student used the ballad of Lizzie Borden as her theme for a stop-frame model animation. The figures were of wire-frame and painted foam, about 16 inches high. This was a major project in her final year.

Digital electronic equipment

As television is a relatively small area of employment for graphic design graduates it is almost impossible for design colleges to find the capital to invest in professional broadcast standard equipment or the peripherals required to support electronic studios. Seeing the potential of training young designers on their own equipment some manufacturers have donated or arranged longer term loans of their machines.

About half a dozen colleges in the UK have Quantel equipment for student use, some since the mid-1980s. The following colleges have Quantel Paintboxes:

Bournemouth and Poole College of Art and Design
Duncan of Jordanstone
Farnborough College of Technology
Manchester Polytechnic (new university status –
    name not confirmed)
Middlesex University
Ravensbourne College of Design and Communication

Ravensbourne, at Elmstead Woods in Kent, is unique in having an electronic production centre, which includes a Quantel 'Harry' and which is available to students in the third year of the BA (Hons) degree course in Visual Communication Design.

The tutor on this moving-image design option is Colin Cheesman, a former Head of Graphic Design at the BBC. He believes that there are great advantages in the unusually well-equipped department. He reports:

Over the past four years, Ravensbourne College of Design and Communication has, with the generous assistance and support of many companies in the television industry, particularly Quantel, developed a specialist course for film and television graphic design and animation. With a purpose-built television centre, equipped to broadcast standard, the College offers professional training to students of programme operations, studio engineering and graphic design.

The School of Graphic Design has access to an electronic graphics suite comprising two Quantel Paintboxes, a Harry digital production system with Encore, video animation equipment and a 16 mm film rostrum camera with editing facilities. All the graphic electronic resources are linked to the studios and post production areas in the television school.

In their second year graphic design students have an opportunity to discover, through an introductory animation project, if they have an aptitude to think and design with moving image graphics. Selected students are offered the option to make film and television graphics their major area of study. They are expected by the end of the course to be thoroughly conversant with the operation and technical capability of the various types of available equipment, but this is never at the expense of the uppermost objective of the course, which retains the pursuit of intelligent creative design.

Design and programme operation students are able to work together on 'live' projects. These provide a realistic working experience with 'on-air' deadlines and the discipline of working as a team member. Further advantages in inter-school co-operation are evident when design projects require special sequences of live-action. Camera and sound crews from television are keen to assist on these projects that can entail location or studio recording. Time management and scheduling of work is also regarded as one of the essential basic skills to be acquired during the course.

Both the college and Colin Cheesman are very much aware that changes in television technology are swift and continuous. However, providing just a few of the next generation of graphic designers with hands-on experience, as well as a strong understanding of some of the equipment available in industry, and the ability to make design judgements, is a significant contribution.

The current mood

Are those who are at present complaining about the lack of resources available for art and design education right to blame government policy, or is it now unreasonable to complain at a time of obvious severe economic trouble? Should the critics acknowledge that a period of change designed to bring improvements (e.g. the independence bestowed on the polytechnics/universities) is bound to be unsettling now, and that the medium and long-term future is bright?

Patrick Heron's forthright article in the *Guardian* in November 1991 made reference to the 'absolute mayhem which government has continued, with deadly consistency, to wreak on this nation's utterly brilliant and world famous art schools'. They were once, he argued, 'autonomous art schools, which used to swim with visiting painters, sculptors, designers of international repute, all dropping in as "visiting tutors"'.

Heron regrets that art colleges have become minor sections of faculties within very much larger institutions. For example, what was once an independent 'School of Art' is now likely to be a 'School of Design and Media' within a 'Faculty of Law, Languages and Communication' within a 'University'.

The chain of responsibility concerning decisions about capital expenditure on specialised materials, equipment, or future building necessary for new processes have become very remote to those in power. They can also become extremely divisive when other faculties are competing for the same pool of funds.

The art historian Professor Sir Ernst Gombrich, in his book *Topics of Our Time*, published in November 1991, gives this admonition:

> Those who hold the purse strings are fond of repeating that 'He who pays the piper calls the tune'. Let them never forget that in a *society* wholly devoted to practical skills there can be no pipers and that those who call the tune will be met by uncomprehending silence. And once the pipers are gone, they may never be heard again.

Interest in the continuance and development of training has tried to keep pace. Many organisations have been, and still are, deeply involved outside the efforts of the broadcasters themselves: The BKSTS, The Royal Television Society, the CITE, the television industry trade union BECTU (formerly the ACTT, which has a number of 'accredited colleges') and BAFTA. The full names and addresses of these and other organisations are listed at the back of this book.

Training within ITV

Prior to the creation of Channel 4 there were only three regimes in which television graphic designers could usually work – the BBC, the IBA contractors, or, much less likely, among the tiny band of freelance designers who worked mostly on television commercials.

In ITV there were union agreements with all employers (both large and small) and these controlled entry into the television industry, which for some years did not expand and had very few jobs. This and other factors resulted in newcomers being only the most single-minded types of individual. Once they got 'inside' they did not leave. There was nowhere to go, using the same job skills, other than the BBC or abroad. Once in a graphic design department their promotion was likely to be very slow. Supervision was extremely close and no one progressed in the jealously protected hierarchy of this unspoken 'apprenticeship' scheme without becoming very adept at their craft. The results were high standards but a coterie which tended to protect information and be rather insular. The BBC system assigned an assistant to every senior graphic designer. For a senior designer to have an assistant was not a luxury; it was the best method of learning yet devised, as most older graphic designers will attest.

Graphic design training at the BBC

Richard Higgs is one of the few graphic designers to have worked for both ITV and BBC (An outline of his career is given at the end of this chapter.) He has contributed his opinion of training within the corporation:

> To become a graphic designer in the BBC you need to have studied to a BA degree level or an equivalent and have a good working knowledge of design, drawing, typography, photography, the printing processes and video production.
>
> In the past students were taken on as holiday relief at assistant level and given the chance to work with a trained designer. Following this period they would be encouraged to apply for a post as 'Assistant Designer'.
>
> After a period of between four to six years the assistant would be given more responsibility and would be expected to apply for the grade of Graphic Designer. During this period they would have become conversant with television design practices including storyboarding, collaborating in the design

deformations and deflections, however, not only depends on the amount of loads but also on the ways in which they are applied to the structural part.

Although its object is materials and structures other than human subjects and relationships, mechanical theories may help advance psychosocial studies with not just the terminology but also the ways of thinking and modes of study. Particularly, for our concern with the practical issue of unidimensionalization in constructing composite measures of multidimensional psychosocial constructs, there are so-called strength theories that would render very helpful clues.

In predicting the "well-being" state or measuring the relative strength of structural parts of various materials, strength theories specifically deal with the issue of unidimensionalization. The mechanical properties of structural materials are normally determined by tests which subject the specimen to comparatively simple stress conditions, whereas the strength of materials under more complicated cases has only been investigated in a few exceptional cases (Timoshenko & Young, 1962). Various strength theories therefore have been concerned with relating the simple criteria of allowable stresses to the complicated conditions which occur in practical design. As Timoshenko and Young (ibid., p.314) states, "The purpose of these theories is to predict failure conditions under combined stresses, assuming that the behavior in a simple tension or compression test is known."

The first theory is called the maximum stress theory, considering the maximum (in terms of absolute value) principal stress as the criterion for strength. This is the simplest form of theoretical assumptions underlying the process of unidimensionalization of the measurement because all other dimensions next in importance are simply omitted. There are many examples that are not amenable to or even contradict the maximum stress theory (Timoshenko & Young, 1962). However, this theory served as a milestone indicating an effort of scientific reduction based on a thorough analysis or "factorization" of complicated stress conditions as well as an explicit theoretical assumption that lends itself to empirical falsification. It was a momentous departure from arbitrarily choosing one or more individual indicators, which brought the art of measurement to a scientific standing.

The second theory is the maximum strain theory. The hypothesis is that a material fails when the maximum (in terms of absolute value) equals the failure point strain in simple stress test. As there are well-established theories like Hooke's Law relating strain with stress while stresses of all directions can contribute to the strain along one specific dimension, this theory represents a better

of all types of work, preparing quotations, completing the project on time and within budget, commissioning specialist contributors, from animators, model-makers, and illustrators and be involved in the commissioning and recording of any music or soundtrack required. They would also have gained working knowledge of studio production, working with film and VTR editors in electronic edit suites and would be using both traditional and electronic design equipment.

The television industry is very small and in the BBC people are encouraged to move around in order to gain experience. This is done by applying for short-term attachments to a different department, or even location, within the company. If this temporary change was successful they would be offered a permanent job.

This type of process has been eroded as the industry has become casualized. There are fewer staff posts and a greater number of employees on short-term contracts. It is now not uncommon for a department to have over half the designers as temporary staff. In the ITV companies there has traditionally been less training. They were able to recruit from 'defecting' BBC staff, and in the present climate there is a shrinking workforce as companies make redundancies in all areas.

With the impending 'Producer's Choice' policy within the BBC, and the cost-cutting in both the BBC and ITV network, the expectation of employing assistant designers is diminishing. A programme producer will not wish to pay for a designer with an assistant if it is felt that the cost of 'training' will come directly from their programme budget.

Although there are many small production companies being started, as the producers are being encouraged to 'go independent', these will never be able to offer the range and diversity of training that the BBC and ITV have been able to do in the past.

Another ex BBC observer, A. J.'Mitch' Mitchell, a great contributor to post-production video effects (he advised on the renowned commercial for Anchor Butter's swaying cows and is now on the board at Moving Picture Company), contributed to a Royal Television Society conference on training in 1989 as follows:

The BBC was large and financed in a way which allowed it to have well-organised training programmes spreading from its London base to the regions. There was, as stated, a good relationship between the new recruits and the already established senior design staff. More formal teaching was also available in training units. The corporation always attracted high talent and there was soon a seepage of well-trained staff to other parts of the television world which the BBC had trained but could not always employ at higher levels.

This flow of creativity provided about half the senior staff of the rest of the industry over the years, and is a major reason for the consistency and uniformly high quality of the whole of the British broadcasting and film industry.

The title of Mitchell's well-argued paper (Television *The Journal of the Royal Television Society*, July/August 1989) was 'How little training can we get away with?'

Deregulation and training

News of deregulation in the broadcast system has made every section of television begin to campaign for more effort to train the next generation in the industry as thoroughly as in the past. Conferences like the RTS enterprise in 1989, 'Training for survival', are a part of that concern.

Much has already happened. The long 'apprenticeship' of ITV is doomed to end. In their preparations to face the challenge of more competition many of the ITV contractors – certainly those who have lost their franchises – have already begun to shed large numbers of their most thoroughly experienced staff. The six-month initiation into all departments given to BBC newcomers, to allow them to see how each worked, no longer exists in the slimmer organisation. Of course the newly appointed companies will begin to recruit, but they are setting up at a period when advertising revenues are collapsing. As there is fierce rivalry for each market share they will tend to employ from the recently filled 'pool' of redundant BBC and ITV staff. They will also rely on the freelance system and not recruit permanent staff. No problems? Unless the pool dries-up!

Costing graphic work at the BBC has gone through a number of phases. Soon, under a new freedom already mentioned, 'Producer's Choice', programme-makers will be able to buy the services they require from any outside supplier. Unless the internal departments' costs, including those in graphic design, are lower than, or competitive with, the outside markets they are in danger of being abolished.

All this may produce economies and very balanced accounts. Where will the responsibility and the cash for training come from? Has anyone made a plan or even a prediction? Mitch Mitchell's lecture concluded:

> Creative opportunities offered by the new technologies are fantastic, but without people trained in the art and craft of television and storytelling then they are surely doomed. It is reasonably straightforward to teach people about a particular machine or even a technique. It is not easy to teach people to have feel for the medium as a whole. To do this they need years in a TV studio or post-production facility.

Work experience

Large number of careers have been shaped by the work placements that colleges or students themselves have managed to arrange. This is a vital part of the training process *for both sides*. BBC Centre, their regional stations, all the ITV stations past and present, and many facility companies and design groups – all of these have been willing to take part in this two-way process. Some of the CVs below make the value of work experience dramatically clear.

Professional liaison

Very many television stations, facility companies, post-production houses and design groups have given advice and encouragement to students attempting to enter the industry. Some, through college staff, have given practical assistance by allowing students to use equipment in 'down-time' to carry out animation and other technical processes – from making soundtracks to using digital effects devices for their college projects. The Mill in London has helped students on the graphic design course at Ravensbourne in this way. Most of the independent design groups all over the country have co-operated in similar ways. They have also toured the art colleges with showreels, to try to portray some of the pains and pleasures of television graphics.

How have people entered the profession?

In all professions it is too easy to categorise people and stereotype their career profiles. Thankfully there are many different ways of joining and of contributing to the world of television graphics. Beyond the overriding element of chance we can be guided by our knowledge of others as role models, observing their successes and mistakes.

Routes towards the relatively small target of television graphic design have been very diverse. From the explorations of young designers leaving college and seeking employment, to those who can now report on a lifetime in television, here are some of the pathways:

Nine graphic routes

1 Peter Ming Wong: Close encounter of a digital kind

Peter Ming Wong started his full-time art education on a BTEC two-year course in the School of Art and Design at Blackburn College. His National Diploma work gained him acceptance on a three-year BA (Hons) course in Graphic Design at Bath College of Higher Education. This course is registered with the CSD (Chartered Society of Designers) to allow qualifying students to apply for diploma membership when they qualify. Like many thousands of his generation Peter had used a home computer (a Spectrum) for play. Then in his second year at Bath he concentrated on the study of 'electronic media' with a very small group of students selected from the 30 in his year.

The syllabus covered: desk top publishing – this was backed by access to Apple Macintosh machines – and he gained confidence in designing with computers by producing two-dimensional images and text for print; video animation – this had a video rostrum camera using low-band recording linked to a three-machine editing suite. He was able to gain experience with simple stop-frame model sequences. No paint system was available at that time.

Two external projects gave him more self-confidence. In one he designed storyboards and was able to shoot a 15-minute live-action video to promote the services of a local uniform manufacturer.

His final year major project led him to conceive a storyboard which needed computer animation beyond the scope of the college equipment. A breakthrough came when his personal tutor, Mike Grey, suggested he might work with postgraduate student Nigel John, who was working on a PhD in Computer Science at Bath University and wanted an animation problem to program. Peter wanted to realise his storyboard. A perfect equation – art and mathematics.

The making of an epic

Five months later and after many hours of work *Hit the Road Mac* was premiered, using Nigel John's software. The low-cost equipment had taken up to three to four hours to render each frame of the 64-second animation.

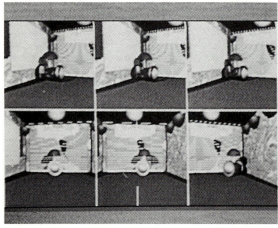

Photographs of models made by Peter Ming Wong when on an HND course at Bath College, to assist in digitising the figures for computer animation. Below, are some of the many frames he had to make displayed on the monitor as they were rendered

It was not exactly *Gone with the Wind*, but it was a considerable achievement (enhanced with synchronised soundtrack by courtesy of HTV).

The software was advanced enough to provide 3D modelling, colour-mapping, reflections and shadows. Peter's storyboard was analysed and amended at weekly meetings. He made 10 inch-high models of the robot performers to aid the digitising process and show their proportions.

The project benefited from the humorous theme of the storyline – a number of colourful robots are anxious to see the Bath College Degree Show, when a faulty blue automaton sadly collapses and fails to arrive.

Peter so enjoyed this encounter and the learning process involved that he decided to apply to two CNAA validated MA courses which stood out in the field of 'computing in design'. He was accepted by both Coventry and Middlesex Polytechnics. (There is also an MA course in Image Synthesis and Computer Animation at Middlesex intended for trained animators and designers who wish to extend into computer work.)

In his year at Middlesex Peter Ming Wong found the instruction and his introduction to new equipment of great value. The hardware was a Spaceward Supernova – this is an interactive device providing two- and three-D animation where the user has direct control of the menus without addressing a program to get results. About 60 per cent of the students on the course were design biased – graphic imaging, textiles and animation; the remaining 40 per cent were architects, interior designers, or maths-based programmers. Both sections were required to learn at least one computer language. Peter got to grips with PASCAL and used the PICASO (sic) and PRISMA libraries of digital formulas.

Using their chosen language the students learned to address information via keyboards at terminals connected to the polytechnic's mainframe. This allowed them to construct objects taken from their storyboards and then create in-between frames for animation. The mainframe connected to the frame buffers, in this case ICON, and the individual frames were rendered in colour on monitors. At this stage Peter found the more powerful equipment reduced the rendering time to about 30 to 40 minutes per frame.

His second adventure into the computer landscape – the story of *Splat the Fly* – confirmed Peter's real talent as a designer. He started the sequence with live-action video where a young boy is playing with a computer where he has to swat flies. When he reaches a huge score of dead flies a remaining single fly seeks revenge. It manages to escape from behind the monitor and succeeds in swatting him. The final shot shows a real fly resting on the keyboard.

While working on a short-term contract in the Graphic Design Department at HTV in Bristol, Peter proved that his chosen training had equipped him to be extremely capable and versatile, enabling him to adapt quickly. There he learnt to operate their paint system, Aurora, working on general programmes and the news.

He set out on a visit to the USA and Canada at the beginning of 1992, modestly describing his 'little computer knowledge' as an 'attempt to understand'. 'I do not

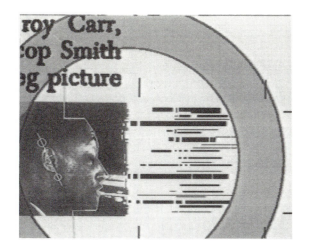

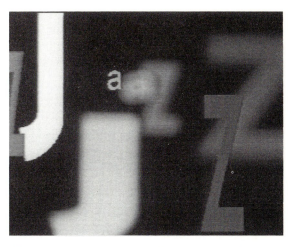

Most graphic design courses will encourage students to investigate their own personal interests as much as is practical. Animation, in its many forms, is certainly one of these

Middlesex University has facilities for film and video work and these are two examples of student animation projects. The stills above are from an animation made by Jason Keely in 1991 as his entry in a competition for a TV commercial to promote an all-jazz radio station. Steve West made a short film whist at Middlesex entitled Private View *for which he made a giant-sized version of the familiar 'pin-screen executive toy'. This was shot in stop frame on 16 mm film and it received many showings at film festivals throughout the world – including Los Angeles, London and Canada. He has since then been commissioned by MTV to make a 10-second channel promotion based on the same theme*

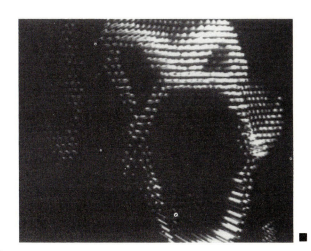

want to be an operator but it is vital to know what electronics can achieve', were his parting words.

Wherever he decides to settle he was clear in his determination to be creative.

2 Kieran Walsh: A transatlantic launch

Kieran Walsh describes his compact experience of industrial learning, which he has achieved in a time of recession, both in this country and the USA:

For a graphic designer, getting into television can be far from easy. Where do you begin? The route I took is probably the most common.

Upon finishing my A levels, I took a one-year foundation course at Hugh Baird College, Merseyside. Though a general art and design course it was especially strong on graphic design which happily suited me. The college prides itself on getting a very high percentage of its students on to the higher education course of their choice. The consequent pressure exerted the choice between working like never before or leaving. This nine-month period was less than enjoyable but it helps considerably in teaching self- motivation, an invaluable quality for a BA course when you are really left to sink or swim.

In 1988 I embarked upon a BA (Hons) course in Graphic Design at Bristol Polytechnic, my first choice. The first year was quite rigidly structured, providing a good overview of design and illustration. In Year Two most of the course direction is relinquished to the student. At this point I was developing a strong interest in design for television and I was able to tailor it to my own needs. Now that the course is modular this is even more the case.

I began to arrange extended work placements, first at BBC Bristol, then at HTV West. Later I went to Indiana University of Pennsylvania on the Polytechnic's exchange programme. IUP has its own cable television station on campus. This has industrial sponsorship to furnish it with broadcast quality equipment. The idea is to provide experience in all aspects of production, from vision mixing to autocue. The four months were worthwhile, but I would not advise on such a course in isolation. It is vital to learn a specific 'craft' such as design and then find out about the industry once you are 'in'.

In my third year I sent off more letters and CVs than I care to remember. Making contacts before

you leave helps to avoid the 'post graduation rush'. One of a myriad of letters to the USA paid off. Aurora Systems, manufacturers of high-end paint and animation work-stations, invited me to intern with them in San Francisco. Having used Aurora during my work at HTV I already had a certain amount of knowledge but in the US I received several months of intense training from very talented people. (Students can obtain a temporary work visa through the British Universities North American Club.)

Six months out of college I now freelance between HTV in Bristol and Aurora in the USA, getting a wide variety of design tasks from them.

So, how do you get into television design? In most countries including Britain a BA is usually expected. I like to think talent helps. But overall it's not luck, more a positive, persevering attitude: 'There are no losers, only winners who gave up too soon', someone once said to me . . . and I'm sure there's some truth in it.

3 Doug Foster: Motion and emotion

Doug Foster is a camera operator and lighting director at the London-based company Cell Animation. His reputation has brought him to work with some of the most imaginative and demanding graphic designers. Here he describes the emotions and responses that led him to his present world.

I think I understand graphic designers in television because I come from a similar background. I did a BA in Graphic Design and Animation at Liverpool Polytechnic – a really good animation course run by Ray Fields. Sue Young, the animator, was one of his students.

As an assistant at the Open University – which was my first job at the BBC – I got to know the way in which they work, the budgets they have, and the way they think. I soon became more interested in models and lighting effects, as doing flat animated work with the rostrum camera can mean dropping cels one-by-one until the early hours. So I ended up becoming an assistant cameraman – a post no one wanted – and soon became acting cameraman. On requesting an attachment I joined a marvellous man, Peter Willis, at the London Television Centre for five weeks and learned much more.

The young department at Milton Keynes got into the routine of producing very interesting opening titles for routine programmes – amazing 'Bladerunner-rip-offs' – often using models. These would cut to the presenter lecturing in a 70s suit and kipper tie, on 'Thermo-nuclear Dynamic Mechanics'. I felt I was not in the right place!

I thought I would try to build a motion-control camera. We were already shooting models as best we could on an ordinary film rostrum by removing the rotating bench. Using 'Dexion' we operated a mirror beneath the table and simulated a motion-control rig. All this was operated with a BBC microcomputer, for which I got someone to write a program to control the rig – all very amateur but we learnt a lot! The camera was a Neilsen-Hordell but even with endoscopic lens and periscopes, when we could afford to hire them, it was all too limiting.

Next I went to John Cook, a manager in BBC Graphic Design in London, to show him my configuration for a motion rig. He listened but reported that it would be two years at least before motion-control in the regions would be possible and even then they were not certain of its value.

Luckily I then heard about the new rig at Cell Animation. They were a film company with seven conventional rostrums at that time, under their Managing Director, Paul O'Hagan, and they had the same problem that I had – they were attempting to shoot 3D model work with these!

One job at that point was a version of *The South Bank Show* for Pat Gavin – it featured an animation through the Mona Lisa's eye and down a piano keyboard – and all this was achieved using models on the old rostrum. It was much more advanced than my Open University work and I thought 'Here is something really worth working for'.

I had my showreel of all those flashy little titles from the OU and my Silver D&AD – shared with Rob Kelly, now a graphic designer at Lambie-Nairn. The piece was *'Manufacturing Systems'* and it was made with scratch-cuts and multiple exposures on a minuscule budget of £100.

O'Hagan took me on as the first camera operator to work on their new Mark Roberts ceiling-mounted computer-controlled motion-rig with an S35 Mitchell film camera. Very soon afterwards Cell Animation purchased one of the first Quantel 'Harrys'.

Doug Foster's work in animation occurs where creative design and technical knowledge are inseparable.

---

4 Liz Friedman: Wide horizons

Successful graphic designers at the BBC can enjoy a rich diet of very stimulating work. After the expected apprenticeship as an assistant graphic designer and working for BBC News when she first left Hornsey College of Art, Liz Friedman has produced titles for many prestigious programmes.

Last year she designed new titles for *40 Minutes* and *Horizon*. Other programmes include the drama series *The Dorothy L. Sayers Mysteries* and *Lord Peter Wimsey*, and *Around the World in Eighty Days* which starred Michael Palin. Much of this work has given her a string of prizes as well as, she says, 'A great deal of pleasure – much of that thorough co-operation with excellent technicians of all types.'

She has lectured at Humberside, Norwich, the London College of Printing, Middlesex Polytechnic and Ravensbourne, as well as the Online Computer Graphics convention at Wembley Conference Centre.

---

5 Colin Cheesman: Head of Graphic Design at the BBC and director of design company

Colin Cheesman was rare among art students of his generation in having a passion to work in film or television. He seems unsure where the desire to combine his art school training at Hornsey School of Art with the moving image actually arose.

His first design job was in a West End advertising agency, Colman Prentice and Varley, but his first move towards the small screen was when he was hired as a junior in the art department of Pearl and Dean. This gave him the valued trade union membership of ACCT. After 18 months he gained a post with the then very young ATV in London. Working first near Oxford Circus he later spent four years learning as much as he could as an Assistant Graphic Designer at Elstree Studios.

In 1963 he started to work for BBC2 and when he left the corporation in the late 1970s he had been Head of their graphic design department, having guided it through important years of its evolution.

With Bernard Lodge he enjoyed the challenge and complexity of running Lodge Cheesman at a time when there were few rival independent graphic design

groups. Since that partnership ended Cheesman has worked as a private consultant and lectured at the major training colleges for graphic designers for television – Norwich, Kingston and Ravensbourne.

6 Martin Foster and the graphic Olympics

Martin Foster was a designer/animator at Electric Image when he described his route to that point in his career:

> At Kingston I specialised in animation for my last year and a half. It was what I always wanted to do. In 1980 while still at College I won the Royal Society of Art Bursary in television. The subject was *The Olympic Games* and four years later I was working at the BBC on presentation graphics for the real games. By 1988 I was here at Electric Image producing the 3D computer work for the BBC animated titles for the Olympics yet again! During my Kingston years I was encouraged by Jean Braid as a tutor, and met a number of BBC graphic designers who set us briefs – Graham McCallum, Rik Markell and Michael Graham-Smith among them.

Martin Foster's 'sporting life' in graphic design started, as so many of his contemporaries, working as a holiday relief at the BBC Graphic Design Department at Television Centre. After three months he attended a formal interview board and was then offered a permanent post as an assistant graphic designer.

I was set to work in Presentation. It was a marvellous place to work at that period – a training ground – ten really talented graphic designers all dividing and sharing work and ideas. Everyone seemed to help everyone else. Oliver Elmes was head of the group. I was an assistant to Graham McCallum. Assigning an assistant to the designers was not a luxury – it was the best way of training. I was there for a year and a half before I went to work as one of two designers on the Horizon *programme*.

Later Martin joined the main graphic design programme area and finally began to design titles . . .

> I did the up-date of the classic Roger Ferris titles for *The Old Grey Whistle Test* . His were 10 years old then and were the longest running of any programme, I believe. I had only three weeks – such a rush. The series had been offered to other designers and they had all managed to argue that there was insufficient time! I was an assistant and the brief was sitting on a desk when I rushed in and said 'I want to do this'. It was my life's ambition.
>
> Within hours I presented my storyboard. A shop dummy in a smart suit watching a TV set. We would have some lasers flashing about. I raced around getting all that together – bought a dummy, had the hands and head chromed, got a suit and arranged the shoot at Damson Studios. It was the first film shot there. The cameraman was Douglas Adamson. The music was by Dave Stewart of Eurythmics and the reason for the change of title? 'Old' and 'Grey'

Martin Foster's model film animation for The Whistle Test *was designed fairly soon after he joined the main Graphic Design Department at the BBC. He took on the difficult task of replacing the very long-running title, designed some ten years previously, by Roger Ferrin. Foster's interests later led him to 3D computer-aided design at Electric Image*

News programmes and election broadcasts have been a major factor in the wider introduction of electronic graphic equipment to other programme areas. Here (right) is a still from an animation for a BBC election programme with electrically generated type and computer animation

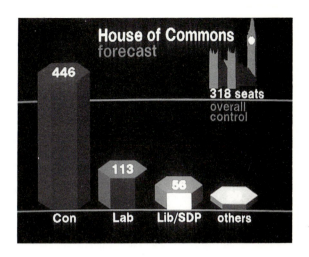

7 Hazel Dormer: Graphic news

were to be dropped and 'jazzed-up' to *The Whistle Test* .

I had been at the BBC for six years when I happened to make a title for the Wimbledon Tennis Championships whilst working with a sports group headed by Darrell Pockett. My storyboard was taken to Paul Doherty at Electric Image where the estimate of £12,000 for the computer animation was out of the question at that time, so the sequence was produced as best we could on the in-house computer-controlled rostrum camera for £300! The OB producer refused to use the animation but my discussions with Electric Image drew me back to them and they offered me a job. And, yes, the first assignment here? Handling the computer animation of those BBC Olympic Games titles!

I had done a one week course at Middlesex Polytechnic with Dr John Vince using Picaso on their mainframe and had a strong response to designing with computer images. At school I came from a science background and did Maths and Physics at A level. I have found that very useful in this part of my working life. I can talk with mathematicians and programmers, understanding their language, and with designers, becoming a bridge – many graphic designers are still very perturbed by technology and some programmers would not know red from green.

For me computer animation has been a marvellous interlude yet I will return to design.

Computer processing is destined to become ever faster and more interactive with pull-down menus with no keyboards and better software for character animation. The future will continue to surprise.

Channel 4 has a news service separate from any other group. The headquarters of Channel 4 News are in the City Road and the graphic design team is headed by Hazel Dormer. Here she describes her route to her present post:

I studied for my BA (Hons) degree for three years at Newcastle-upon-Tyne Polytechnic. In my final year I studied computer graphics on a 'Spaceward' computer system under the lecturer Peter Leake.

In my spare time I worked voluntarily at the BBC Newcastle. After college I applied for, and got, a 'runners' job at ITN in London, involving getting slides for use on graphic transmission and checking Quantel Paintbox discs for the designers. In this time I taught myself how to operate Paintbox and after six months I gained a designer's job on *News at One*. Two years later I became a senior graphic designer on *Channel 4 News* and then the head of graphics on that programme one year later. I have just designed the *Channel 4 News General Election* style and am having an exhausting time in the run-up to the election.

8 Richard Higgs: An indirect approach

After gaining his diploma in graphic design Richard spent five years in London as a freelance designer and illustrator. Then his interest in fine art, drawing and painting became so great that he decided to return to art school. Accepted at Norwich School of Art, he immersed himself in three more years, diverting into animation and scenic design. His BA thesis turned to Television Aesthetics and introduced him to many ITV and BBC programme makers, in particular to Robert Harris, a BBC set designer.

Graduation led him to short-term contract work for the BBC in Norwich on their then newly installed computer-graphic system, Spaceward. He also taught on the foundation course at the School of Art in the city before his next move towards graphic design for television – a three-year contract with the BBC in Leeds, unusually combining graphic design *and* set design for programmes of many kinds. This took him to a post of staff graphic designer for BBC Manchester where Richard designed network programmes – *The Travel Show Guides, World Embassy Snooker Championships* and two series of Janet Street-Porter's *Rough Guides* for *DEF II* .

In 1989 he moved to Independent Television as a Senior Graphic Designer working for HTV Bristol. In a small department he revels in his direct involvement in every aspect of the company's output – from shift work on regional news and current affairs to title design for all and any programme – and 'hands-on' work with 'Harry', and 'Aurora', HTV Bristol's electronic paint and 3D workstation.

A personally acquired Macintosh IIfx has recently become part of Richard Higgs's world, used he says as 'a design tool for work and for personal projects'. In 'The Reel World' (Chapter 7) he generously describes very frankly his experience of TV graphics on minuscule budgets.

---

9 Jerry Hibbert: Animator and graphic collaborator

When Jerry Hibbert was a young student at Ealing School of Art one of the part-time lecturers was a graphic designer at the BBC. His need for help with simple in-betweening led to Jerry 'volunteering'. This led to evening classes in animation and an introduction to a 16 mm rostrum camera. His final year animation project got him the post of trainee animator with George Dunning who was then working as the director on the Beatles-inspired *Yellow Submarine* (1968). By 1974 he had begun to work as a freelance animator. His association with animation for television graphics came

through a long-lasting and close collaboration with Pat Gavin.

Work for Gavin and London Weekend Television included the titles for *Two's Company*, which starred Elaine Strich and Donald Sinden, and award-winning animations for the various *The South Bank Show* titles.

Jerry Hibbert set up in partnership and is now a director of Hibbert Ralph Animation, one of the most progressive companies of its type in this country.

Footnote to education

Paul Rand, one of the most inventive art directors on the American scene of the post-war period, has made an eloquent contribution to the debate on design education in his book *The Designer's Art* (Yale University Press, 1985). He notes the absence of a well-formulated or systematised body of literature on the teaching of the subject of design, making the problem more perplexing than it might be:

> The subject is further complicated by the elusive and personal nature of art . . . the problem still remains how best to arouse the student's curiosity, hold attention and engage their creative faculties.

Through an admitted process of trial and error he resolved that the instinct of *play* was the key element and quotes Gilbert Highet (*The Art of Teaching* New York,1950), to substantiate this view:

> The best Renaissance teachers . . . spurred their pupils on by a number of appeals to the play principle. They made games out of the chore of learning difficult subjects – Montaigne's father, for instance, started him in Greek by writing the letters and the easiest words on playing cards and inventing a game to play with them.

Rand concludes that many of the psychological and intellectual factors implicit in game playing are essential to successful problem solving, e.g.

curiosity	skill	excitement
anticipation	perception	enjoyment
challenge	observation	discovery
promise	analysis	reward
	improvisation	

Rand's hypothesis is that problems set with clearly defined rules as well as an obvious sense of play are those most likely to succeed. The best professional work underlies the effectiveness of this philosophy. □

Chapter 6   Tools of the Trade

When is technology 'new'?

There are considerable myths about the application of electronic technology. These range from over-enthusiasm where computers are credited with solving everything, to technophobia where designers have a distrust of the whole movement. A main concern among designers is the alienation believed to be caused by the computer's assumed dehumanising effects.

Technical change has been, and always will be, a source of genuine apprehension in every activity, posing moral and practical problems at every level. Artists have fought over them – often with great passion.

Occasionally those who protest about change fail to see that what they cling to was once revolutionary technical innovation. The discovery of photography to record the visual world became 'a powerful stimulus to artists, setting them off on a quest for new means of expression' (*The Camera*, Life Library of Photography, 1970). In a parallel manner, the long chain of closely related discoveries affecting television image-making has had many stimulating consequences.

Animators who favour traditional hand-drawn animation (using film, cels, painting and tracing) and are reluctant to use computer-aided animation systems could overlook the consummate technology required to develop and build a well-made film rostrum camera, without which their drawings could never be filmed.

There are designers who have certainly over-estimated the contribution of the computer. No one should have the illusion that computers can draw (well or even badly) by themselves. They do not create or design.

Recently admired works, like John Lasseter's computer-based animations for Lucasfilm, *Luxo Jr*, *Red's Dream* and *Knickknack*, and Nick Park's stop-frame film *Creature Comforts*, for Aardman Animation are praised essentially for their visual humour, not for their technical prowess, which was in both cases superlative. Creativity, storytelling and imagination should always predominate.

Most techniques seem to survive the alternating periods of fashion and neglect.

Genuinely useful products and ideas are absorbed to serve alongside 'the new' in a continuous evolution and in endlessly resourceful ways. Recently there has been a strong reaffirmation of the essential contribution of hand-animators in the production line of computer-aided sequences. Animators spend years observing how things move and how to represent movement by drawing. Their contribution was misguidedly underestimated for some while.

Artists *and* technocrats?

How should graphic designers in the late twentieth century deal with television technology?

The work and attitudes of two artists with very diverse temperaments and styles, Norman McLaren and Eduardo Paolozzi, prompt answers to this question. Both have proclaimed themselves 'pro-technologists'.

McLaren, a man of great inventiveness in the world of film animation, born only 20 years after the birth of the moving image (Lumière brothers, Paris, 28 December 1895), left his studies at Glasgow School of Art for the lure of the new-found technologies of film-making and, later, animation. Paradoxically, once his energies drew him towards the techniques of animation, he strove to break down the barriers that the film rostrum camera seemed to impose upon him. He succeeded in drawing directly on to film in black and white as well as colour, fulfilling his wish to get as close as a painter might to a canvas, yet all the time he was seeking his own technical innovations. *Stars and Stripes*, 1939, and *Boogie Doodle*, 1940, among others, are testaments to this early achievement giving what computers have bestowed so much later – cameraless animation. He even managed to make percussive and semi-musical sounds in *Dots* and *Loops* (both made in 1940) by marking the optical soundtrack area adjacent to the picture frames on the 35 mm film, rather than use

case of unidimensionalization. Nevertheless, there are also many cases in which the maximum strain theory may be shown to be invalid. Especially, experiments show that homogeneous materials under uniform compression can withstand much higher stresses and remain elastic (ibid.).

Because of all the failure cases of the first two theories, there came the third, the maximum shear theory. "This theory assumes that yielding begins when the maximum shear stress in the material becomes equal to the maximum shear stress at the yield point in a simple tension test" (ibid., p.315). Since the maximum shear stress in the material is known to be equal to half the difference between the maximum and minimum principal stresses, and since the maximum shear stress is equal to half the normal stress in a tension test, the total difference between the maximum and minimum principal stresses becomes the criterion for strength (ibid., p.316). This turns out to be a better example of unidimensionalizing measurement, as the theory is in good agreement with experiments and is widely used in machine design, especially for ductile materials (ibid.).

In addition to these classical strength theories, succeeding development has included some newer theoretical approaches to strength of materials, such as the distortion energy theory which considers the strain energy of deformation per unit volume of the material as a basis of selecting working stresses in machine design. Timoshenko and Young (1962) compared some different strength theories. It is noticeable that the difference between results of different theories is considerable. This indicates the significance of theory in scaling in general and in unidimensionalization in particular.

The most general state of stress which can exist in a mechanical body is always completely determined by specifying three principal stresses. In behavioral and social sciences, however, the situation can be far more complicated: The number of dimensions we have to deal with is too often larger than three, which do not lend themselves to spatial visualization and thus usually can only exist in our abstract reasoning. Besides, the human body and mind are distinct from the material part studied in mechanics. However, as we have borrowed the ideas of stress, support, etc. from the mechanical world, there are a number of useful lessons and inspirations we can draw from the above review. First, and foremost, it shows that there is a real possibility to unidimensionalize the results of multidimensional scaling in a scientific fashion. And there are different pathways to achieve this objective by means of alternative theoretical models. Second, appropriate theoretical assumptions are the key to unidimensionalization. The appropriateness of theory can be directly tested by observational data. Third, specificity is the basis

recorded music. He made 60 films but none of them had any speech. He was so devoted to animating images to music that he said he could not incorporate words in any way – 'I couldn't do it. It would be an intrusion of a very alien kind.'

The National Film Board of Canada holds the collection of the meticulously kept technical notes McLaren prepared on his animation experiments from 1934 until 1983 when he retired. This quote comes from a two-page report on his methods of making *Rythmetic*, 1956, an amusing film using animated numerals:

> The numerals appear and disappear with very short 10-frame fade-ins and fade-outs. These fades were not done with the camera; instead, for each numeral, we made a series of ten replaceable cut-outs, in progressively darkening shades of grey. This method of fading involved less labour than using the camera for the great number of fades that we had to do.

He recorded similar details on his method of *drawing* the soundtrack. The combination of the attributes of artist and technician, of creativity and scientific study, are obvious. Shared research of this calibre, between scientists and artists, has been the essence of computer graphic conferences like SIGGRAPH and PARIGRAPH.

In a film interview during the mid-1980s McLaren said:

> If there is technical invention a person with an artist's nature is likely to take that thing and use it. We are getting young artists using computers to make a new kind of animation and a new type of film. That is only right and proper. That's the way it should be. I know that if I had been growing up now I would have gone right in and tried to get hold of a computer and tried to do things with it.

And he added:

> If one is a child of one's age – born at a certain time – one uses certain things.

Man and machine

Paolozzi was one of the creators of, and a contributor to, an influential exhibition entitled *Man, Machine and Motion* held in the summer of 1955 first in Newcastle-upon-Tyne and then at the Institute of Contemporary

Reproduced by the kind permission of the National Film Board of Canada

Norman McLaren grasped the available technology of 1930's film animation with great delight but throughout his long career he constantly sought ways of adding to his chosen medium. He drew and painted directly on to film; he created sounds by making marks on the film's optical soundtrack; and he left meticulous notes describing, he said, 'what at the time, seemed to be relatively innovative about each film's technique'. Above, he is seen marking a soundtrack in the 1950s and (below) is work in progress on Rythmetic. *The strip of film (left) shows the hand-drawn scratches on* Loops *(1940) and (far right) are part of McLaren's lucid notes on the making of* Rythmetic *(1956). McLaren epitomised the amalgam of creative artist and technologist required in graphic animation*

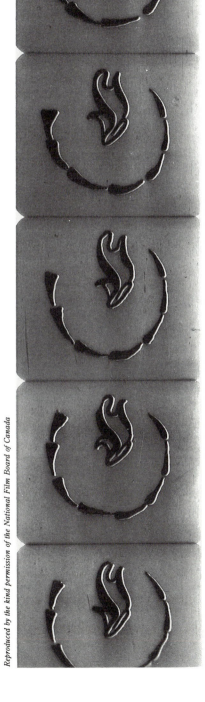

The numerals (about 1½" or 4 cm high) were cut out of stiff white paper, and animated on a large black card on the animation table-top.

For precise positioning of the numerals we drew lines and marks on the card using a dark red pencil, well sharpened to give a very fine line. Because of the fineness of the line and the darkness of the red on the black, the camera did not pick up the lines, but they were clearly visible to the animators who were relying on them for proper line-up, and moving of the numerals.

Such dark red marks, in the shape of small ticks, permitted preplotting of the callibrations of much of the animation.

The basic cut-outs of the numerals were rigid, in other words, we could not animate their internal form, only slide them around the field.

To get internal flexibility we used two methods. With numerals 4, 7 and 5 we jointed their elements where the dots occur:

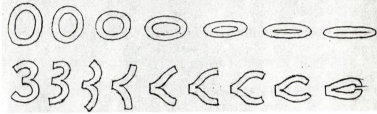

With numerals such as zero and 3, we made a series of replaceable cut-outs:

This 'music' was made by engraving small marks on the sound track area of black emulsion-coated 35 mm film, with a knife, razor blade and stylus or needle. Broad knife-strokes gave deeper pitches, very thin needle-scratches gave high pitches. Strokes going fully across the sound track were loud; strokes going only partially across the track were quieter. The degree of quietness could also be controlled by how diagonal a stroke was. Strokes at right-angles to the track had maximum volume; the more sloping a stroke was to the track the less volume it gave.[1]

Many of the peeps clicks, clacks, clucks and thuds consisted of only a single scratched stroke on every fifth or tenth or twentieth frame. Percussive sounds that had a more definite pitch consisted of several scratched lines closely grouped together.

Examples of scratching on sound track (enlarged):

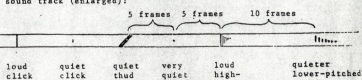

			5 frames	5 frames	10 frames
loud click	quiet click	quiet thud	very quiet click	loud high-pitched sound	quieter lower-pitched sound

This directly engraved original track served as a master to rerecord from. During the rerecording slight amounts of reverb were added here and there. Because of the rerecording, the appearance of the track on release prints is that of standard variable area.

Art in London which showed his interest in over 200 machines 'which extend the power of the human body'.

Throughout his career, Paolozzi has sought one medium after another, often restlessly and imaginatively combining them. He engages technologies old and new, and tries to exploit all of them. He will carve or scratch into plaster and then print conventionally on to paper; he finds and constructs forms, interlocking them at will in very unusual ways. These have then been reproduced using traditional methods of metal casting to make a sculpture. No technique seems foreign to him. He has said 'I feel that man's position in general is reflected in his feeling and attitude towards the machine. You can retreat from it or you can wrestle with it.'

Graphic designers in television have certainly had to wrestle with it! They must react continuously to one process after another and, in the manner of McLaren and Paolozzi, not only employ new techniques but, wherever possible, improve on what they find. Many designers become adept at all of them and as a body their creative hopes continue to inspire progress.

No designer can ignore the engineering structures of television, however complex, and at the very least they must be able to converse with understanding.

The politics of technological change

The pace of the introduction of innovative electronic equipment was certainly not determined by the manufacturers. Management, trade unions, the designers, the engineers and technical staff, the programme-makers and the advertising agencies have each played some part in promoting or retarding its acquisition. No one group has taken a consistent attitude – they have all been 'pro' and 'anti', at various stages, to suit their own purposes.

There have been times when managements have held back for financial or technical reasons. The unions have resisted new technology when their members saw that the rate of change from an established system might affect a whole generation of skills. On other occasions they have pressed management to move with the times.

Graphic designers have appeared similarly ambivalent – some hoped to acquire each electronic device immediately it was 'switched on'. In this they were often abetted by programme-makers hoping to gain advantages over rival channels for improved design and presentation. Other graphic designers have tried to resist new methods. The in-fighting has been long and at times surprisingly fierce.

Paint systems, for example, were born into the video ethos around 1980 but it took years for them to be purchased in significant numbers. The reasons are those indicated above, together with the lack of will of those in power to be more enterprising. The first to purchase were often private investors.

Technical standardisation

The quality of television images is determined by the broadcasting signals. The standards for colour television, which affect all designers, are the result of decisions made in the 1950s and 60s. These were made nationally, and the adoption of the incompatible broadcasting standards – PAL, NTSC and SECAM – in various countries has handicapped the development and interchange of equipment and laid the foundations of a labyrinth.

Transmission standards

The first colour transmission system was NTSC (National Television Standards Committee), designed in the USA in the 1950s and now the standard in North America, including Canada, and Japan, giving enormous purchasing power and audience coverage. The images are resolved in 525 lines at 30 frames per second. Its main limitation is its non-stable colour in transmission and reception.

The Europeans sought to improve on this and, although the UK was tempted briefly to use SECAM, the BBC chose PAL (Phase Alternate Line). This German invention (Walter Bruch of Telefunken) was a latecomer with the higher resolution of 625 lines and 25 frames a second.

France made the political decision to go its own way and build a unique format. The French created SECAM – 625 lines and 25 frames per second – which was later the system adopted by Russia and the Eastern bloc states. SECAM is used only for transmission. As more advanced digital effects, editing and videotape recording appeared there were difficulties. Now French studios and facilities have PAL or component equipment encoded to SECAM format for transmission.

Diverse video tape standards

Some five years ago there was hope that the video tape divergence, the so-called 'video format wars', would end with the introduction of the digital era. This has not occurred. There are now three digital formats – 'D' for digital – D1, D2 and D3. The manufacturers Panasonic, Sony and Ampex have recently launched three new ones – Panasonic/D5 (four is considered unlucky in Japan), Sony/Digital Betacam and Ampex/DCT.

Lack of conformity, national and international, has left the problem of incompatibility of basic systems, causing immense confusion.

Technical complexity

A brochure produced by a facility house to *help* their graphic design clients included this paragraph:

> The first problem is the range of video formats. Our equipment works in the 'domains' of digital video (601 and 605), component RGB, component YUV and transcoding RGB machines into YUV, but on the whole, all video paths must be available for reasons of flexibility. This means that video signals have to go through transcoders* of different sorts to connect two pieces of equipment together.

Video signal formats

There are presently three standards in which electronic video equipment is operated in the UK. These are: Composite PAL, Component and Composite Digital. A fourth, Component Digital, is to follow soon.

Graphic designers need to be aware how they are used, and understand their relative merits.

Composite and Component signals are both analogue. That is, they operate on continuous physical quantities. Each time an analogue signal is processed there is a loss of quality and the introduction of 'electrical noise'. It is often said that when digital signals are used there is no loss of quality *no matter how many times a process is copied*. This is true only if there is no bitrate reduction system used.

All colour television picture or computer-generated images consists of three signals – red, green and blue (RGB). Video formats are methods for converting these RGB signals into coded electrical forms that are more efficient than RGB. However, in graphic design production on computer work stations ('Paintbox' etc) RGB signals are used directly (Diagram A).

Composite systems (such as PAL) encode the RGB signals in a 'composite' signal containing both the colour and the luminance information. This may then be transmitted down a single cable or a transmission channel. However, a composite signal is degraded when it is processed electronically (Diagram B).

Component systems encode the RGB signals differently. A component system essentially produces a luminance, (black-and-white) signal and two colour difference signals which are all kept separate during subsequent processing. This ensures much less degradation than that occurring in composite coding (Diagram C).

While broadcast TV stations are still mainly composite, modern studio engineering is being converted to use component systems. At the point of terrestrial transmission even a component system has to produce a composite signal (PAL etc).

In addition, both 'component and composite' equipment is available in digital form to provide all the advantages of digital processing where suitable.

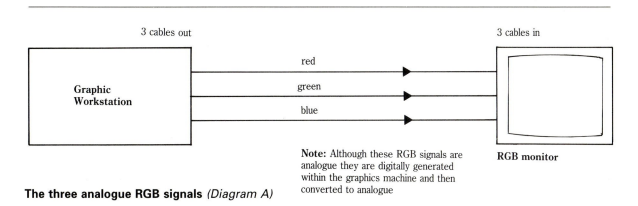

The three analogue RGB signals *(Diagram A)*

Note: Although these RGB signals are analogue they are digitally generated within the graphics machine and then converted to analogue

* When trancoders are introduced they contribute to a lowering of quality as they are not 'transparent'.

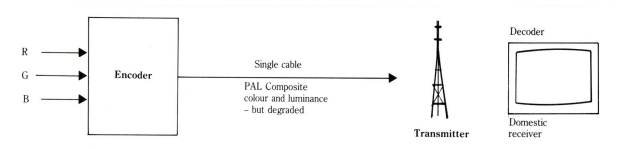

The Composite system (PAL) *(Diagram B)*

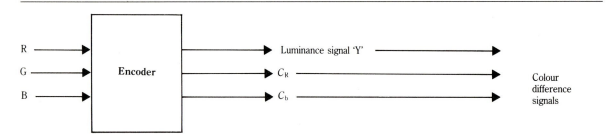

The Component system *(Diagram C)*

Before describing some of the digital video effects equipment (see page 43) that now forms part of graphic design production the diagram below shows how computer image-making devices, like 'Paintbox' and DVE equipment operates.

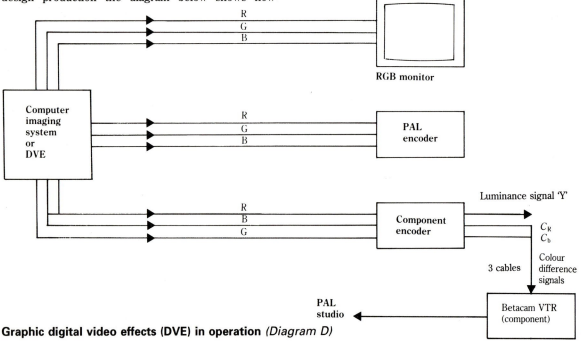

Graphic digital video effects (DVE) in operation *(Diagram D)*

Component and composite compared

Component systems have better signal quality. Why? The luminance signal and the colour difference signals are kept separate.

Composite (e.g. PAL) combines the signals in one signal. Therefore degraded detail and colour can occur.

Three-dimensional graphic animation work is now carried out in RGB digital mode and then transferred to composite prior to transmission.

Video signal formats

(in order of quality of image on-screen)

Format	Description	Usage
1 RGB	Primary RGB signals	In high-definition computer processing when creating graphic images and animated effects
2 Component digital	Digitised component signals	In Betacam digital quality machines D1, D5, DCT and Digital Betacam
3 Component analogue	The image is encoded as three signals: (a) A luminance signal (i.e. black and white) (b) Two separate colour difference signals	In modern TV studio facilities (except transmission) and in Betacam SP equipment
4 Composite digital	Digitised PAL with the advantages of digital storage and manipulation – no image degradation	Drop-in replacement machines (especially VTRs) in the composite environment D2 & D3
5 Composite analogue	Any signal described as PAL, SECAM, or NTSC is a composite signal. Colour and luminance are encoded as one signal	In broadcast TV one-inch C-format VTRS

The format CCIR 601 is a de-facto standard for component digital television signals which is specified as the 4:2:2 sampling system.

Digital effects devices brought to television graphic designers the means to manipulate the moving video image at the touch of a button. This power was not always employed wisely. The examples (below) from an Abekas A51 show a single frame from two typical routines. They are 'picture explosion' and 'cubemaker' – where all four separate picture inserts are animated. DVE machines work in RGB format and diagram, (D), (opposite) shows how their images are displayed while operating on an RGB monitor and are then converted to PAL for transmission as well as to the Component format for VTR editing

Video recording and storage standards

These standards are more recently established than those in film and they are in a less resolved state, hence more complex to describe. The future may see them refined and, hopefully, reduced in number.

Video *tape* standards (in descending order of quality)	Usage
1 **Digital tape D1, D2, and D3**	Highest standard in post-production
2 **Half-inch tape** Betacam SP (Super Performance) M2 (Panasonic) Analogue/component	UK Broadcast standard
3 **One-inch tape** C-format	Library transmission and international exchange standard. Most broadcasters use half-inch for transmission
4 **Umatic High-band**	Non-broadcast, corporate, educational and showtapes
5 **High 8 and S/VHS*** (Sometimes used for broadcasting at reduced quality)	Where normal formats are impractical or not available
6 **VHS/and S/VHS*** (Same basic quality as High 8)	For viewing work and domestic standard

* Graphic designers now carry about their 'work-in-progress' on these disc formats.

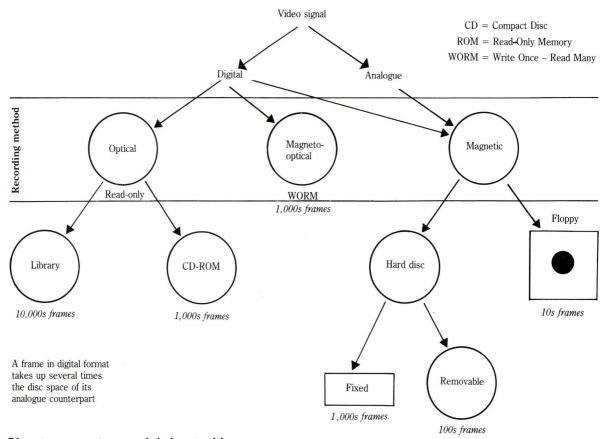

CD = Compact Disc
ROM = Read-Only Memory
WORM = Write Once – Read Many

A frame in digital format takes up several times the disc space of its analogue counterpart

Disc storage systems and their capacities

How can graphic tools be classified?

Electronic equipment to produce on-screen graphics is operated by the broadcasters and by the facility houses that have grown so rapidly to support all television programme production. Three divisions help to classify the mass of devices which are described as hardware or more colloquially 'kit'. Descriptions of equipment made in the television press and in the publicity of the facility houses are usually under the following headings.

The first is *editing equipment*, the second *digital effects devices*, the third is *graphics equipment*. A summary, far from exhaustive, of some of the items most widely used in the UK and Europe is given below.

The term 'editing' is self-evident in this context but it can be carried out in a 'linear' fashion or, to the advantage of graphic designers, by 'non-linear' methods with the 'Harry' family. *DVE devices* are those which manipulate, 'spin, flip, and tumble' as the manufacturers say, still and animated pictures.

Graphics equipment defines those machines which create the primary images for stills or animation.

Typical equipment in graphic production

1 Editing equipment

Linear	VTR machines: e.g. Sony, Ampex Operated in editing suites using various formats: D1, D2, D3, one-inch, BetacamSP and M2
Non-linear	Quantel's range: 'Harry' 'Flash Harry' 'Harriet' 'Henry'
	'Avid' 'Lightworks'

2 Digital effects devices (DVEs)

e.g. 'Abekas' A53, A51,
'ADO' 100,1000, 2000, 3000
(Ampex Digital Optics)

'Carousel'	'Kaleidoscope'
'Charisma'	'Merlin'
'ME5000'	'Mirage'
'Encore'	'Myriad-fx'
'Gemini III'	'Solo'
'Zeno'	

3 Graphic equipment

Primary image producing for stills, animation and lettering

Digital paint systems

e.g. 'Artfile' 'Supernova'
'Videopaint' 'Venice Paint'
'Matador' 'Vertigo 190'
Quantel's 'Paintbox':
V, VE and EVS series & Classic

Paint systems with 3D or animation

e.g. 'Aurora' 'Symbolics'
'Matisse' with 'Rodin' Software
'Pastiche'

3D animation systems

e.g. 'Alias'	'FGS Elite'
'AVA3'	'Iris'
'Bosch FGS4000'	'Pixel'
'DGS'	

Software (3D computer animation)

e.g. 'Softimage' 'Wavefront'

Motion-control rigs

e.g. 'Made-to-measure' by
Mark Roberts Nielsen-Hordell

Character generators

e.g. Ampex 'AVA' Aston 'III and IV'
Aston 'Motif'
Quantel's 'Compositor'
'Chyron'
'Cypher Sport' & scoring keypad

4 Stills stores and libraries

e.g. 'Gallery 2000'/Logica
'Picturebox'/Quantel
'Slidefile' Rank Cintel
'Wallet'/Aston

5 Other services

Telecine
Film editing for all standards
Transfers from film to video and video to film/all standards
Film duplication
Conversions from one video standard to any other:
PAL, NTSC, and SECAM

The tradenames listed are representative of only some of the vast range of the well-known equipment in the industry at this time.)

Equipment in the UK facility houses

These have the reputation of being the best equipped and most versatile in the world. The high standards of graphic presentation set by the television industry in the UK have encouraged facility companies to invest and prosper. Among those favoured for their services to graphic design are CAL Videographics, Carlton, Complete, The Frame Store, The Mill, Moliniare, The Moving Picture Company, Rushes, SCV Television, Tele-Cine Cell and TVi.

Television standards and tools – from BC to AD

The study of the equipment that graphic designers in television have used for image-making, animation and sequence editing can be divided into two periods. In 50 years of transmission the first 35 years might be regarded as 'BC', before-computers – while the past 15 are the period 'AD', or aided-design:

> Between 1975 and 1981, computer technology changed so profoundly that those few years mark a watershed not just in the history of computers but in modern culture as a whole (*Computer Basics, Time-Life,* 1989).

These dates are poignant in relation to television graphics. Prior to 1975 there was hardly any computer-based equipment in television graphics. Since then, and markedly from the early 1980s, there has been a deluge.

Graphic equipment before the computer ('BC')

Here is a list of graphic design production devices and systems in the approximate order in which they came into the hands of the designers, not necessarily the order in which they were first invented. Only the first item* can truly be considered obsolete.

Equipment	Purpose	Images produced
Masseely press*	Proofing type	White type on caption cards or rollers
Rostrum cameras 16 and 35 mm	Single-frame film animation	All cel-based and multiple single-frame drawing animation. Multi-plane. Clik-frame using photographs. Single-frame to real-time still sequences
Aerial-image (An accessory to the film rostrum camera)	To combine two separate film sources	Typically, cartoon figures on live-action film backgrounds
Steenbeck	For viewing 16 and 35 mm film rushes at various stages of production. (A young graphic designer mis-heard this name and spent some time enquiring about a 'Steambed')	

Movieola	Synchronising film and sound in animation	
Stop-frame film cameras	Model animation	The filming of any 3D scene at any scale
	Time-lapse effects	A plant growing in five seconds! Or any similar illusion
Basic film cameras and lenses (Plus use of endoscopic, fisheye and wide-angle lenses, snorkel and periscope attachments)	Live-action	Limitless
Specialised cameras Gyroscopic	Aerial filming	Steady movement
Light weight	Attaching to remote controlled model aircraft etc	Pictures from places otherwise inaccessible
High-speed film cameras	Slow motion images	Typically a bullet in flight or a splash
Under-water camera equipment		
35 mm slides (Colour transparencies)	For captions etc. Now input, via telecine scanners, into DPS or still stores	Stills in colour or in black and white
Telecine slide scanners	Transmitting, or transferring, 35 mm slides	
Telecine machines	To transmit, or to transfer, moving film to video	
CSO	(Colour separation overlay) This permits a number of layers of television images to be combined and transmitted as one picture and it has been essential to graphic presentation for many years, e.g. news presenters in front of graphic images. The principle is used in 'Harry' editing. Subjects are separated by single colour backgrounds, most frequently a bright blue and the system is called 'CSO' at the BBC. Elsewhere there are 'Chromakey' and 'Ultimatte', a later computer-base keying system.	

Film animation 'BC'– before computers

Film standards are relatively simple. They are 8 mm, 16 mm, 35 mm, and 70 mm. The amateur stock known as Super8 has been relegated to the museum by the domestic portable video system, but facilities for viewing and transferring archive 8 mm will always be required. (Super8 was used by graphic designer Harry Dorrington of English and Pockett as a special-effect camera for the programme *O1* as recently as 1990 and for low-budget work. See Chapter 7, section 3.)

Wide-screen film generally uses 70 mm. Films to be shown on television have to be transferred to video tape for transmission to an appropriate format since Telecine machines are hardly ever used on-line nowadays.

The gauges of 16 and 35 mm are both still used in graphic animation. The wider standard is considerably more expensive at every stage – from stock to projection – but its resolution is very good, twice that of the present 625-line TV transmission, and the use of film is preferred on many occasions by graphic designers for this reason.

The unique visual quality of film encourages graphic designers to continue to work with the medium. This suggests it will not entirely give way to video as enthusiasts may choose to use it inventively with current and future electronic advances.

Video has not yet developed cameras to record at the ultra-high speeds of film which can reach 4,000 frames per second. Oxford Scientific Films Limited at Long Hanborough is a company that continues to work with graphic designers in highly specialised filming.

Film-based animation was for 30 years of television the only method available and the *film rostrum camera* and the *stop-frame animation camera* the basic tools of the trade. In both techniques the essential function is the single-frame exposure to build up multiple images which, with the persistence of vision of the eye, appear continuous.

Thousands of titles were produced using film as the sole medium. Viewing a compilation tape of work from the 1950s to the 1980s (painstakingly preserved by the designers at the BBC) uncovered rich and limitless ways of putting the moving image to work. Endless combinations of fades, dissolves, multiple exposures and superimposition were made within the film camera. The use of film opticals and the skills of film editors were the essence of the craft of all animation for television. Now electronic devices make these easier and there are fewer rostrum camera operators and film editors.

Designers working in television we[re interested] in pushing the film medium as far [as their] limited means. The application of si[mple] graphic animation, avoiding the use o[f the] cel system, reached a peak in the 1960[s and 70s. They are typified by the 'cut-out' animations of Terry Gilliam in *Monty Python's Flying Circus* made with few frames, resulting in very jumpy movement.

Cameraless animation was sought and achieved by Bernard Lodge in his first title sequence made for the *Dr. Who* series in 1963. Later he was involved in further work to extend film animation techniques. (See later under the heading 'Computer-controlled film rostrum cameras' Page 50.)

Aerial image animation

Before Mickey Mouse was born Walt Disney experimented with this idea when he combined human live action, using film of a young girl, with line-drawings in a film called *Alice's Wonderland* as early as 1922! He revived the use of aerial-image projection when his studios made *The Reluctant Dragon* (1941), and *Mary Poppins* (1964): 'For sheer entertainment value a sequence mingling live-action and animation is immense' (*Variety Movie Guide*, Hamlyn, 1991).

The aerial-image device, fixed to the basic rostrum, allows moving film to be projected through a glass aperture on the rostrum bed by means of a 45° mirror and re-filmed frame-by-frame. Transparent cels were interposed on the flat glass of the rostrum to combine drawings, or still photographs, with the live-action. Later the video effect of colour separation overlay (CSO), 'Chromakey' and 'Ultimatte' made this almost obsolete. The recent development of 'Harry' and similar video editing equipment has lessened their application. Computerisation of the aerial-image was achieved to give more control and improve output in the late 1970s.

Stop-frame film cameras

This method is very widely exploited now and generally makes use of models, manipulated clay or Plasticine on wire armatures. *Creature Comforts*, which won an Oscar, and the Electricity Board television commercials (both designed and produced by Aardman Animations) are currently the best-known animations to employ stop-frame, while the Czech master of animation Jan Svankmajer has created many films using the same technique. A full-length stop-frame animation film using

for pursuing generality. Before scales can be unidimensionalized individual items must be analyzed by clarifying their dimensionality. The role of principal stresses in strength theory points to the necessity of factor analysis and/or other methods for multidimensional scaling. Fourth, the task of practical diagnosis is two-fold, requiring the assessment of bio-psycho-social conditions on one hand and the establishment of the criteria (critical points of individual or group failure) on the other. It can be said that in behavioral and social sciences, the latter is both a scientific and a political matter. Methodologically, however, it can be done by relatively simple methods provided that there are appropriate theories linking the results to more complicated cases. Finally, it is not necessary for the mathematical representations of such theories to be very simple, especially when researchers are no longer legitimated to rely on hand operation of routine quantitative transformations in the computer times.

With the insights from science as to how general approaches to global scaling may be formulated, Chen (1997) tries to establish some measurement hypotheses and unfold their corresponding scaling programs for the construction of desired global measures. These can be explained as follows.

Hypothesis 1. The effect of a multidimensional measure is determined by its maximum (in terms of absolute value) component score. This can be called the maximum component (factor) theory for scaling any multidimensional psychosocial construct. If G stands for the score on a global scale and abs(P_i) for the absolute value of the *ith* component score, then the hypothesis can be expressed by the equation

$$G = \max\{abs(P_i)\} \qquad i=1,2,...,n \qquad (1)$$

in which "max{}" is a function used to extract the maximum from the assemblage of absolute values of the components that are numbered n.

It should be pointed out that the "maximum component" in this hypothesis is case-specific, considering the relative magnitude of all the components of a theoretical construct (e.g., social support) to each individual, other than its general distribution among the population that distinguishes between the relative importance of components in explaining the total variance. In fact, if the latter is the case, then we have a different hypothesis.

...odels was made from Kenneth Graham's *Wind in the Willows* by Cosgrove Hall.

Real objects, small-scale models, puppets, Plasticine, clay, or figures with movable armatures, and every possible combination of all of these, are still manipulated in front of the stop-frame camera. Constant light levels and perfect continuity are essential to obtain the fluidity which deceives the eye, and the 'old-fashioned' film camera still produces superb work.

Some studios still specialise in stop-frame filming of models and these are equipped with dollies which permit the easy movement of the mounted camera, although computer motion control has superseded this equipment. (See Chapter 8, section 2.)

Colour separation overlay

This daunting technical tag was used by the BBC to describe the video effect now seen habitually in newscasts and weather forecasts to present graphic information behind the image of the presenter. This system of keying-out parts of the screen has been used creatively by graphic designers, with their set design colleagues, in every type of programme. Thames

Television relied on Ultimatte extensively in their award winning *Una Stravaganza dei Medici*, made in association with the facility company Framestore, and Chromakey is used to display still and moving pictures beside presenters in news bulletins as routine system.

This technique has, at its best, forced the responsibilities of set designer and graphic designer either to be shared more thoughtfully – with more creative results – or it has occasionally led to confusion and conflict between the two.

The principle is simply to record part of the screen against a single key colour (a bright ultramarine blue is the most commonly used, to avoid the range of flesh tints) to overlay the background picture and then to video the insert in a separate signal. When the two signals are overlaid a complete image is presented as a single frame. Both components can be moving. The illusions are limitless.

As the technique was improved the too obvious 'edges' were removed and lack of shadows was overcome, making the effects very convincing. The method introduced in ITV was 'Chromakey', at the BBC it was 'CSO' and a later system is known as 'Ultimatte'.

Large areas of a Thames Studio at Teddington were covered in bright blue material as part of the colour separation overlay process – an essential part of creating many graphic effects involving graphic designers

Equipment after computer assisted design – AD

The tools of the pre-computer age are fairly simple to describe. Those based on the versatile microchip can be very ambivalent. If only each computer-based device had one clear function! Craig Zerouni, a director of ComputerFX and a regular commentator on technology, pinpointed this 'convergence' as follows:

> These kinds of products are converging; the boundary between a paint system and a DVE is blurring at the same time as the edge between 3D systems and paint systems disappears. (*Videographics*, February 1990)

Equipment	Purpose	Images or effect
(In approximate chronological order of application to daily work)		
Video tape editing using VTR (e.g. Sony/ Ampex)	To enable video recording to be 'cut' frame-by-frame in analogue or digital mode. Timecode numbers locate every frame	
Computer-controlled film rostrum cameras (e.g. Oxberry Niellsen-Hordell)	To speed-up and enhance film animation	Typically streak-timing and slit-scan images
Digital video effects devices (e.g.'Mirage', 'Encore', 'Abekas')	To manipulate still and moving on-screen images	Originally simple picture twists and turns – later complex and seemingly 3D animation
Character generators (eg. Aston's 'Motif'and Quantel's 'Compositor')	To produce electronic lettering in the video system	
Digital paint systems (e.g. Quantel's 'Paintbox', 'Flair', and 'Aurora')	These devices allow drawing and painting and photographs to be prepared directly into the video system	Limitless in style or pictorial form
3D computer animation systems (e.g. 'Bosch FGS 4000' and 'Symbolics')	To construct objects that can appear to inhabit and move in three dimensional space	
Computer-controlled motion-rigs (e.g. Mark Roberts)	To move any type of model and allow frame perfect repetition	
Animation control units 'EOS' BAC-900 and AC580/ AC580J	To edit frames recorded on a video rostrum camera	Colour video animated sequences
Quick action recorders	To assist animators with black-and-white line-tests as video frames. In-between drawings can be inserted in any sequence very rapidly	
Computer-controlled video rostrum cameras	To record and animate artwork or to shoot sequences from photographic stills or paintings	Precise moves over flat artwork of any description. Fades and mixes only possible at post-production
Stills stores (ESS) Essential to the processing of all graphic images. Large-scale stores are organised as libraries. (e.g. Quantel's 'Picturebox', Logica's 'Gallery 2000' Rank Cintel's 'Slidefile')		To hold instantly accessible electronic images to be displayed singly or at any other speed
Graphic editing devices (e.g. 'Harry', and 'Henry' 'Avid' 'Lightworks')	To enable complex video editing, mainly for graphic animation, in conjunction with digital paint system to amend each frame	
Video stills units (e.g. Sony, Hitachi Polaroid)	To produce hardcopy prints of any on on-screen image	High-quality colour photographs for reference, storyboards and publicity

Video tape editing

Video tape recording was a reality for a long while before editing was practical. The first crude attempts were done with physical razor cuts to the tape. Until the electronic editing breakthrough no real advance in video graphics could be made. When time code was introduced where every frame is numbered per 25th of a second precise editing was possible with a single-frame controller.

Computer-controlled film rostrum cameras

The computer came into television graphics via the film rostrum camera. This new tool of computer-aided animation was pioneered in the USA.

Robert Abel, working in San Francisco, was one of the earliest to produce notable work. Bernard Lodge's many years as a graphic designer at the BBC led him to work with Filmfex, a London-based rostrum film company that advanced early computer-controlled animation and installed an Oxberry rostrum by the early 1970s.

Until then the routine purpose of the film rostrum had been the patient filming of thousands of painted cels, registered on pegs, placed over backgrounds. This was labour intensive but comparatively simple. Disney studios' success was in making cel animation an efficient factory process for full-length feature films and *Beauty and the Beast* is a revival of that skill. But the early television designers were faced with making extremely complicated movements on all types of artwork (flat top-lit artwork, back-lit negatives or colour transparencies) with the rostrum bed and mounted single-frame camera. This process of filming was both hazardous and slow until the microprocessor intervened.

Computer control of all movements of the apparatus speeded up the shooting and allowed far greater access to animation images; the north-south, east-west positions of the bed and its rotation, the zoom of the camera, the focus and exposure were *all* motorised and could then be programmed from a single keyboard.

This increased control meant that far more complex sequences were affordable. Two clearly definable effects were fashionable for a few years – *slit-scan* and *streak-timing*. Slit-scan produced some beautiful moving images where flat artwork, or back-lit colour transparencies, shot portion-by-portion, literally through a slit mask, could be distended in perspective or in curves. Once digital-effects machines arrived, (e.g. 'Mirage' and

Prior to the use of digital video effects devices, the computer-controlled rostrum camera was one of the ways of creating unusual moving images. This example of 'slit-scanning' came from Lodge/Cheesman, designers who did much to develop this technique

'Encore') to produce these kinetic pictures in real-time, as cones, cubes, cylinders and spheres, the era of the computer-controlled camera was very brief. The 'neon' blur of streak-timing was suitable for some subjects but it too became over-used, often inappropriately.

Digital effects devices

The ability of television system to twist and turn pictures from pre-programmed equipment is used by graphic designers in preparing their animations. The use of 'Encore' and 'Abekas' are referred to in the title animations reviewed in Chapter 7.

Designers have always been wary and sometimes hostile to the programme-maker's reliance on such equipment without design consideration. Purposeless swirling pictures, or strangely inappropriate 'page turns' are just some of the 'defects' that can be obtained all too easily and pass for graphic presentation.

Further evidence of the multifarious nature of digital effects devices can be seen in Grass Valley's 'Kaleidoscope'. This has the expected effects called 'ripple', page turns and rolls, but it has 3D animation with a moveable light source and 'glare'.

When well-used in post-production processes digital effects machines can conjure results which are undetectable to the eye but vital to the end result. It is only in demonstration that their prowess can be seen in multilayering image after image with virtually no loss of picture quality.

Abekas Video Systems produce a range of machines to aid the manipulation of images. These examples are from the Abekas A51 which they describe as a '3D effects machine'.

Graphic Designer
MICHAEL GRAHAM-SMITH
Illustrator
JOANNA ISLES
Modelmaker
JOHN FRIEDLANDER
Cameraman
DOUGLAS ADAMSON

The roller captions (above) would have taken hours to set. Now electronic character generators, like Aston's 'Motif', can set and embellish type at any size and in almost any style in only a few minutes

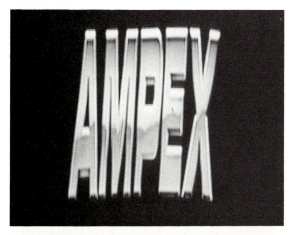

Character generators

Hand-setting and proofing from moveable metal type using hot-metal foil to produce the essential solid white lettering on black card or paper rollers, many feet in length, was finally relegated to the museum.

This late revolution in screen graphics took place only when the battle to produce a satisfactory character generator was won.

Early generators were crude for many reasons. They led a decrease in the choice of typefaces and lowered the standards attained in the archaically slow hot-foil hand-setting method and the other non-electronic alternatives.

The new generators are now an absolute essential in all broadcast graphic services. Their instantaneous reaction makes them perfect for news graphics and they now satisfy the most discriminating typographic standards that can be applied in the television medium.

There is a wide choice but manufacturers have begun to recognise that many of the additional effects and built-in animation, which tended to make their machines more expensive than necessary, were better added through other post-production equipment. Aston Electronic Designs, which has companies in the UK and the USA, are typical in responding to this view and they have produced a completely new series of character generator.

The first is called 'Motif'. This concentrates on the high letter resolution, an attribute vital to typographic qualities and composing needs. Fount characters are presented with the maximum 256 levels of anti-aliasing but the machine has lost the complexity of the three planes of presentation of earlier ranges. Typefaces for 'Motif' come from the extremely wide variety of sources provided by Monotype, Adobe and Font Bureau. On subscribing to a library package (codes within the generator can be 'unlocked' by telephone) access is given to over 1,500 typefaces.

Once presented on the 'Motif' screen any character may be modified. Shadows, colours, slope, weight, inline and outline can all be altered through the roller-ball control and the keyboard. The new character is instantly remembered and a fresh alphabet will be built as each letter is keyed. Lines of type can be set diagonally and there is improved control over the speed of both roll (vertical) and crawl (horizontal) modes. These now have the ability to accelerate and decelerate exponentially, allowing lettering to move smoothly away from, and to, stationary positions. Concentrating on type specification and eliminating the less-needed functions has lowered the cost below that of the previous Aston generators.

Current electronic computer-based character generators have overcome the objections stage-by-stage. An early problem was the definition of lettering outlines where very jagged representation of curves and diagonals spoiled the forms due to the very few picture elements (pixels) available to build up each letter. This was overcome by anti-aliasing, that is adding a layer of pixels – more information – to smooth the rounded forms and the oblique strokes. Screens of the future with high-definition transmission will benefit from graphic and typographic presentation with far higher resolution.

(Paintbox by Quantel Limited)

In the early days preparing stills in the video medium was sometimes referred to as 'painting by numbers'. Now digital paint systems have become the most common tool in graphic design production

Digital paint systems

These are essentially 2D systems for creating still frames using a pressure sensitive pen, or stylus (often with the advantage of being cordless) and an electronic tablet. A large range of 'pens and brushes' of varying widths plus the accepted 16 million colours and every imaginable style of drawing, stencils, cut and paste, and rotoscoping are obtained via a range of menus. Most systems offer typesetting which can be amended with shadows, outlines, etc. Photographs and any other types of artwork can be input by video camera in full colour.

Digital paint systems were the first items to remove the paper, pens and rubber solution from the graphic design studios in television. There is now a wide choice which offers every form of picture-making and re-touching that anyone can imagine. They are essential to picture libraries and to digital editing. Quantel's 'Paintbox' has become the generic term, like 'Hoover' and 'Biro' in their own fields, but there are many others to choose from. Some paint systems have so many features that they are called 'graphics work stations'. 'Aurora' and 'Symbolics' are good examples, as both have programs which enable 3D computer animation to be prepared. In the constant updating of paint systems Electronic Graphics added rotoscoping and cut-out animation to their 'Pastiche' digital paint system – a symptom of all devices tending to converge, as mentioned above.

'Aurora' can be described as a graphic work station as the range is wider than most digital paint systems. Both 2D and 3D animation can be created and displayed in real-time. From from its home in 'Silicon Valley', at Santa Clara it was among the first of its kind. The Atlanta all-news station CNN uses 12 'Aurora' machines for its 24-hour-a-day graphic output.

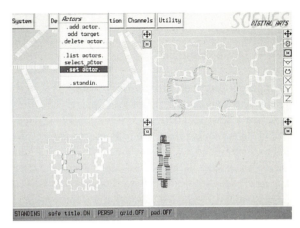

Digital paint systems have become more effective and involved in animation as editing equipment has advanced. At first DPSs were limited to the production of on-screen stills, where the easy combination of colour photographs, graphic elements drawn on the system and type, was instantly valued. Below are a mid-1980s information still from WCBS-TV News of New York designed by Robb Wyatt and a promotion still from an inventive series designed for Channel 4 by Simon Broom

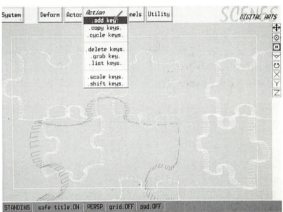

Some computer systems are available to use programs to short-cut 3D animation. Above are on-screen menus of the DGS system in action on an animation for HTV

3D computer animation systems

There are a large number of systems which are largely 'off-the-shelf' machines. They do not require the programming from scratch but are capable of creating 3D moving objects in wireframe, of multilayering images, and then colour and texture mapping these. A device like 'Alias' can do all this, render many frames extremely quickly and feature moving light sources. Bosch produced their 'FGS' in the early 1980s and the 'FGS4500' and other models are still widely used.

A small number of companies use software programs to build 3D animation. One of the most experienced in London is Electric Image and a detailed account of what their work involves is given in section 3 of Chapter 8. The rapid development of 3D computer animation has been presented to the world through the annual SIGGRAPH exhibitions held in various US cities, and the excitement engendered by the delegates – computer scientists, programmers and designers, has to be witnessed to be believed.

Computer-controlled motion-rigs

An account of motion-rigs appears in Chapter 8, section 2. Graphic designers have been using models in their animation armoury since the very beginning. Computers have added a precision to this craft which still depends on the lighting and camera skill of the operators. The Moving Picture Company and SVC Television as well as Cell Animation are three facility houses to offer this valuable service.

The Moving Picture Company in London was one of the first facility houses to offer a computer-controlled motion rig to television graphics. Here their unit is operating with a snorkel lens

Animation control units

These are used mainly by animators rather than graphic designers but many are used in colleges to introduce students on graphic courses to time-based media.

The 'EOS' animation controller pioneered video animation of this kind with a fast single- and multiple-frame full-colour system. There is a broadcast version the 'EOS' BAC900, the EOS580 and a low-cost version called 'Supertoon' for teaching and line testing.

Computer-controlled video rostrums

Video camera brought to animation the advantage of seeing the result of the animation as it was produced. This allowed faults to be seen and corrections made as the work progressed. The final sequence could be transmitted without waiting for hours and the hazards of film baths. Most rostrums, which are nearly all computer-controlled now, are adapted to shoot both film and video. (Continued on page 71.)

Satellite broadcast graphics

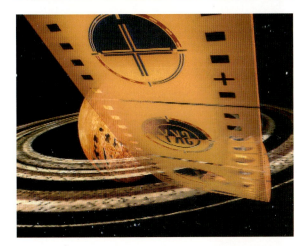

Graphics remain a core service in satellite companies. Mike Hurst, BSkyB's head of graphics, explains in Chapter 2 that most work is made on in-house computer equipment – as was Paul Butler's trailer for Sky Movies *from which these stills were taken*

Independent group design for overseas

TV2, Norway's commercial terrestrial station, commissioned the British group Plume Partners to design their on-screen identity. (Chapter 2) Above – frames of the main identity, a sports version (back of football shirt) and for a weather forecast

On-screen 'typography' ▼

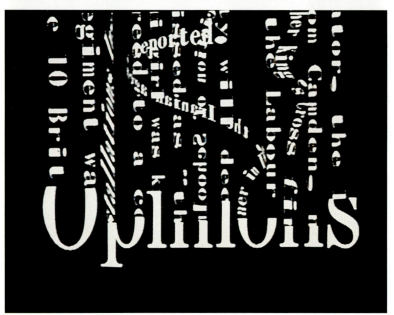

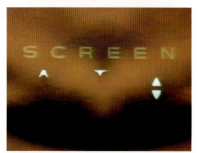

The titles on this page by English & Pockett, which demonstrate the sensitive use of animated lettering, are reviewed in Chapter 4.
Opinions used computer-generated type and gained Bob English D&AD Silver, RTS awards. Animating each letter of the word 'TWO' and subtly echoing the divided screen in the opening frames gave great cohesiveness to Darrell Pockett's title design for the BBC

Encouraging animation studies

(a)

(b)

(c)

(d)

(e)

(f)

(g)

Practical work in animation is fostered at some art colleges. Harrow College is among those with relatively low-cost equipment. The stills show: (a) stop-frame model animation by Sandra Sutton for a sequence on the legend of Lizzie Borden, (b) a frame by Nazfa

Sadougi, and (c to g) a few of the 6,500 images made for an eight-minute prize-winning film Adventure in Barcelona *by Susan Hewitt. More information is given on training for television graphics in Chapter 5.*

Hypothesis 2. The effect of a multidimensional measure is determined by its first component, which represents a maximum portion of its total variance among a population. This can be called the first component (factor) theory. Using the same mathematical denotation as above, we have a general equation to express this hypothesis:

$$G = P_1 \qquad\qquad (2)$$

Compared to the maximum component theory, this hypothesis involves a problem similar to the "ecological fallacy" in research methodology. It emphasizes the relative magnitude of the first component of a measure in light of the total variance among the population explained rather than on its real importance to the individuals, which are in real terms the unit of analysis in most psychosocial studies. Nevertheless, both theories are comparable in that all other components of less importance are simply dropped.

Measures such as these, indeed, may not be regarded as "global." The approaches explicated here, however, are distinguished by clearly articulating theoretically plausible hypotheses and building firmly on empirical components analysis. In applying the second hypothesis, for example, we secure a truly unidimensional scale that fits the data better than any other arbitrarily chosen variables, that explains or represents the largest portion of the total variance or measurement information. Also, the first hypothesis introduces a way of unidimensionalization with more theoretical underpinning by learning from science, though the qualitative difference among various psychosocial components may prevent its use in multifactorial scaling from matching its counterpart in engineering mechanics.

Hypothesis 3. The effect of a multidimensional measure is determined by the sum total of its components. This can be called the summated factors (components) theory, which is the simplest additivity hypothesis applied to components. The mathematical expression of this hypothesis is

$$G = \sum P_i \qquad i=1,2,...,n \qquad\qquad (3)$$

where $\sum P_i$ represents summation of all the extracted components that are numbered n.

Computer animation to MA level

(a)

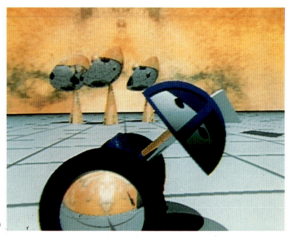
(b)

Various routes towards a career in graphic design are discussed in Chapter 5. Among them is the path of Peter Ming Wong. The stills (a and b) are from Hit the Road Mac *made on his BA Graphic Design course when he* collaborated *with a Bath University PhD student working in Computer Science. Those below (c to f) are from* Splat the Fly, *designed & produced on his MA Course, 'Computing in Design' at Middlesex University*

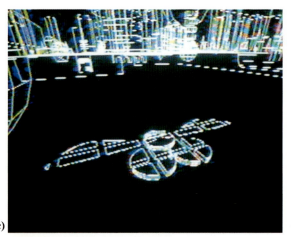
(c)

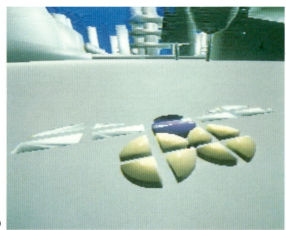
(d)

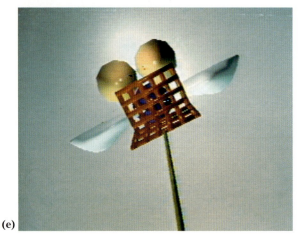
(e)

(f)

Early training in the television industry

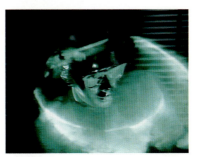

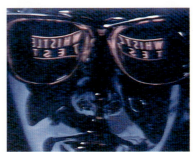

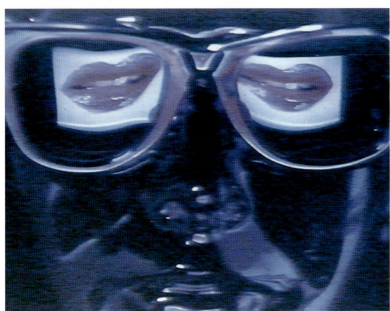

The only way to absorb knowledge of the techniques of animation, which are now more varied and complex than ever, is by working as an assistant alongside senior designers. Martin Foster praises the system he worked under while at the BBC. He broke through to full responsibility as a graphic designer with this title sequence for Whistle Test *when the severe challenge to re-design one of the longest-running series titles came into the department and he was still a relative newcomer. The story of Martin's career, from his student days at Kingston Polytechnic, to his 'apprenticeship' years at the BBC, and the making of* Whistle Test *is told in Chapter 5. How he then contributed 3D computer animation at Electric Image as described in Chapter 8*

3D animation with interactive devices

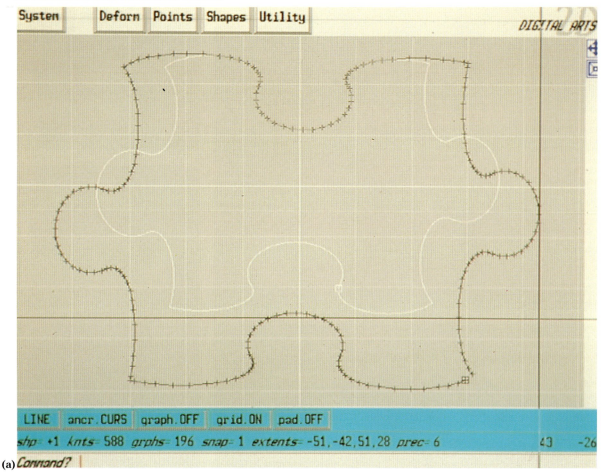

(a)

An amazing array of electronic tools are now available to construct the thousands of frames required for even the shortest 3D animation. The illustrations on this spread demonstrate the application of the DGS 3D system in making the opening titles for an HTV series at first called The Second Imperative *then retitled* The Sexual Imperative. *In this frame* (**a**) 200 points are used to plot the 2D outline of a jigsaw piece which will give a smooth appearance when viewed closely and in (**b**) the lettering is being cut out and built into the jigsaw. Here (**c**) the texture can be seen only half-way around the sphere. In (**d**) the texture is being placed on to a spread-out sphere. Wireframe display (**e**) is a frame of the animation and (**f**) is a test rendering of this in a de-fault red. Then test renderings (**g and h**) show the designer the images which appear on the front and back of the 3D jigsaw shapes. The designer, David James, explains more fully in 'The Reel World' Chapter 7

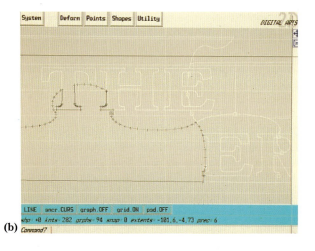

(b)

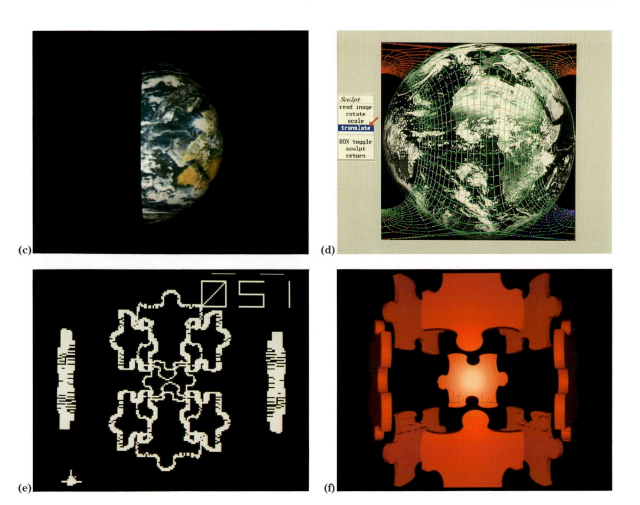

(c)

(d)

(e)

(f)

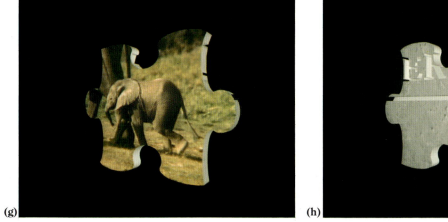

(g)

(h)

(a)

Video editing with Quantel's 'Flash Harry'

A graphic designer working for Quantel, Paul Rigg, produced this diagram (a) to expand his explanation given in Chapter 6, that random access is 'Flash Harry's' hidden power. The monitor menus control the system with a cordless pen (b). The menu on the monitor (1) is the 'desk' menu – with three reels of clips – like 35 mm film. The 'desk' menu (c) shows the pink symbols used to scroll through the clips at variable speeds. At (2) editing a Chromakey sequence is shown – left to right, the original blue-screen shot, then a background kitchen scene, then the combined view, processed through 'Harry's' keyer. The pen is making a 'cut'. With the essential link through 'Paintbox', any frame can be instantly modified. Represented at (3) are the 7,500 full-resolution frames (five-minutes of video) stored on the hard discs. These 'read or write' in parallel, giving the random access and the processing

(b)

which are the real power, while (4) reveals that frames can be selected in any order and as often as required. This non-linear editing gives greater freedom than any previous system and is much faster to operate

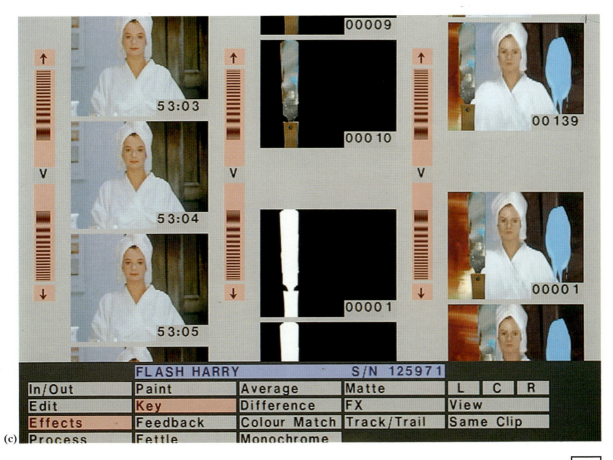

(c)

Providing video lettering

(a)

(b)

(c)

(d)

Character generators were among the first electronic-based tools to be established in television graphic design production. They are now essential to most programme-making. Current developments in their performance, as typified by Aston's 'Motif', concentrate more on 'pure typography' and composition and less on the animation of letterforms which is best handled by digital effects devices. High resolution, a massive range of typefaces, and great finesse in word and letterspacing show the return to basic principles. A roller-ball control presents the menus and some are shown above (**c and d**). Through these displays the selection of type, sizing, opacity, choice of colour and all functions are made. Adaptations to weights, outlines and angle of setting can be achieved.

More details appear in Chapter 6

Rostrum cameras – the enduring tool

The innovative animator Norman McLaren, with his collaborator Evelyn Lambart, (top) using cut-out animation for Rythmetic *(1956). Non-cel film rostrum work was a mainstay for television graphic designers for over thirty years. (See Chapter 6)*

Electronic stills store (ESS)

(a)

Graphic designers working on news programmes are clearly beneficiaries of stills stores. Logica's 'Gallery' (a) presents multiple pictures on-screen (b) which can be searched rapidly through a 'picture request' (c). Other aspects of ESS are reported upon in Chapter 6

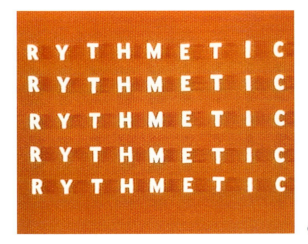

(b)

(c)

Digital paint systems (DPS)

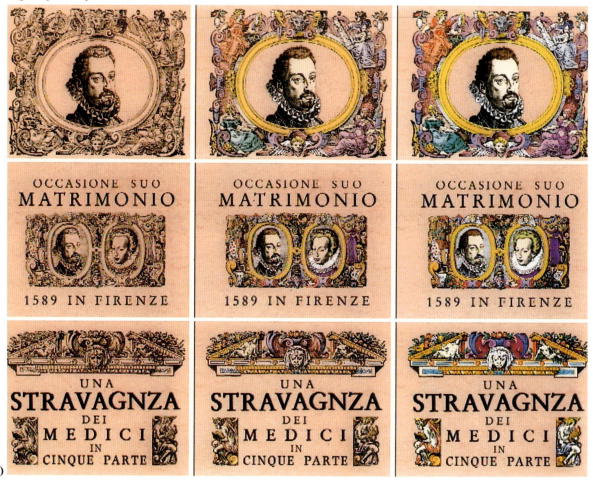

(a)

*The speed and convenience of electronic paint systems
are demonstrated in these test pieces. In tests for* Una
Stravaganza dei Medici *(a) the line engravings (viewed
here as one frame) were input by camera and copied
very quickly. The background colour and tone were then
chosen from the system's palette – in minutes. Painting
with 'watercolour', heightening with opaque white then
comparing results is simple. No other system could
produce the multiple images of the watch (b) so
efficiently. It was placed under a video camera and all
the 12 adjusted shots recorded in seconds. Using
photoprinting, paints and paper for either of these tasks
would take hours; amendments might be irreversible.
Here the original and all the intermediate stages are
stored unaltered. No wonder such systems are used for
storyboards! 'Quantel Paintbox' images/Thames
Television and Middlesex University*

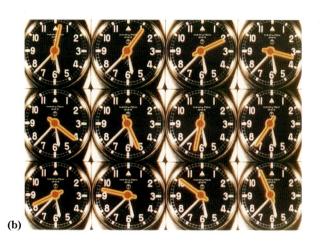

(b)

Motion-control rigs

(a)

(b)

(c)

Liz Friedman describes her design of the titles for the BBC series Chronicle *in section 1 of Chapter 6. Here the stills reveal the operation of a computer-controlled motion- rig. The camera, (in this case using film but video can be used) is mounted on multi- pivoting armatures whose most fractionally small movements can be controlled and repeated as is often required, with absolute precision. Realism is not always the objective but here the qualities of the model and the lighting made the first part of the sequence (**a to c**) very convincing. This complete model was over 14 feet long. Motion-rigs by various manufacturers, each with individual features, have been installed in facility companies – among them The Mill, SVC Television and The Moving Picture Company. Doug Foster shot* Chronicle *and he describes his work on the project in Chapter 8*

(d)

students will set up and proceed with their project in a very orderly manner while others may encounter tremendous difficulties. Until they get to that point, many would not know how they have or have not prepared themselves for the full responsibility of a research project.

Now, let's talk about *you*. Although you are supposed to be guided by a faculty member (or several faculty members), the thesis/dissertation project is completely your own. And the liability will mainly, if not solely, fall on you. If you could start paving the way for your thesis/dissertation research sooner, you would possibly prevent some of the difficult problems from happening when you formally enter the thesis/dissertation stage. As a matter of fact, even with a relatively heavy coursework load, there are steps that you can take to approach your research project earlier.

The study plan

Your educational pursuit in an academic or professional degree program may well be compared to a major construction or engineering project. Such a project cannot go without a general blueprint. Likewise, a general study plan is very important to your success in your educational undertaking. Although most academic/professional programs have carefully designed curricula for their students, you need to work out an individual plan that tailors both academic requirements and resources to your own needs and strengths. Some programs, especially those at the doctoral level, would require every student to formally submit an individual study plan at an early stage of their study (e.g., following their first year qualifying examination). If your program does not require that, you should still carefully think over the major steps that are ahead of you.

To work out an appropriate study plan you must first have a vision of some long-term career direction for your future. You have already oriented yourself by choosing and attending the program. Although you may be reoriented someday, your current study plan has to be grounded on what you can decide at this moment. Based on your career needs and goals, you will need to:

(1) Select a concentration, or a specialty for your advanced study within the program. This will not necessarily set the scope or topic for your thesis/ dissertation. Nevertheless, you are specifying a direction for your study by favoring certain faculty interests, short-listing your course selections, and focusing your attention on a few related areas. It is likely that your thesis or

This hypothesis theorizes the usual practice of summated scaling, except that here it is based on the results of factor or components analysis rather than on the original individual items. In theory, this has several advantages. First, component scores are standardized based on the variance and covariance matrices, which renders an ideal initial base for weighting while makes unnecessary any arbitrary treatment of the original items. Second, it makes clear multidimensionality of the psychosocial phenomena to be scaled and the necessity of supplying appropriate grounds for performing unidimensionalization in measurement. This focal issue is consolidated with the weighting problem in determining the relative importance of various components. As mentioned earlier, it is injurious to completely ignore this issue even in unidimensional scaling. The matter has doubled significance in multifactorial scaling.

It should be pointed out that in the summated treatment of factors (components), a procedure of component scaling could be performed to meet the requirement for additivity. This is similar to the item scaling procedure in unidimensional summated scale development (Donald & Ware, 1984), which is also applicable to the simple summation choice to obtain subscales in place of factor scores in multidimensional scaling. The additive hypothesis entails linearizing the relationships between the scaling items and some criterion variables. Such criterion variables are useful in preparing individual items for unidimensional summation. When component scores are used in multidimensional scaling, this procedure becomes unnecessary since the extraction of components is the break-up of the total variance into orthogonal elements via linear regression.

The pursuit of an appropriate weighting scheme for the purpose of unidimensionalization, however, is extremely difficult in the psychosocial field where the subject matter is highly complicated while the state of the art is yet premature. For example, we do not yet have a theory in life stress or social support studies that has the potential comparable to Hooke's Law in science and engineering, by which we can go some distance beyond the primitive and possibly inaccurate hypotheses described in the above. We do not have any other established transformation formula to relate different components as simple as the maximum share stress theory or as complex as the distortion energy theory in engineering mechanics either. Indeed, in psychosocial studies we are uncertain about the relationship between different components of a specific phenomenon, in terms of their various effects on human life. Without such knowledge, although we can conceive of some mathematically more elegant transformations, such as letting

2 *The Universe Within* – 3D computer graphics

An account of NHK's *medical science series by its Art Director, Kazuo Sasaki, is in Chapter 7. Thirty minutes of landscapes and models of the imagined functions of the human body was an ideal realm for the 3D computer technology in this programme. Illustrations show: a multiple-frame view of the animation in progress (**a**), and a full frame of the final rendering as transmitted (**b**). The sketch (**c**) is one of thousands drawn 'over-and-over again' for the medical consultants' approval and to provide instructions to the animators, and (**d**) shows the final rendering and lighting. An original program, created by High Tech Lab. (employed on this series by* NHK) *– 'Digital Dynamation System', (**e**), enables the 'Wavefront' program to make very soft, flexible movements and this was used in these immune cell scenes.*

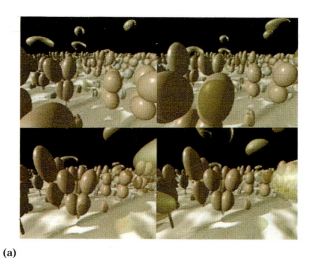

(a)

Credit/Kaneko/Hi Tech Lab, Japan

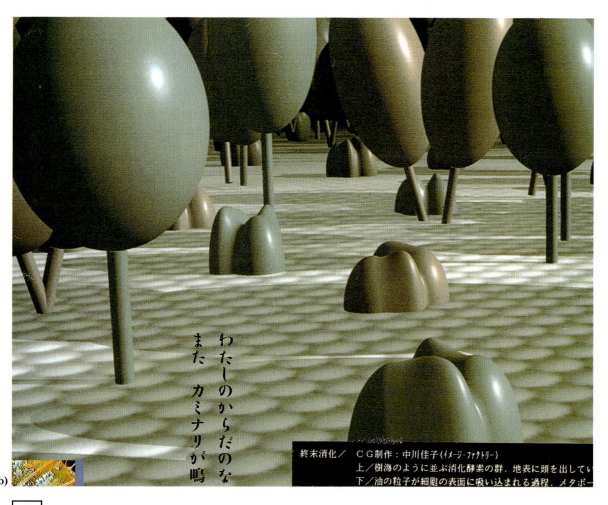

(b)

わたしのからだのな
また カミナリが 鳴

終末消化／ 　ＣＧ制作：中川佳子〈イメージ・ファクトリー〉
上／樹海のように並ぶ消化酵素の群．地表に頭を出してい
下／油の粒子が細胞の表面に吸い込まれる過程．メタボー

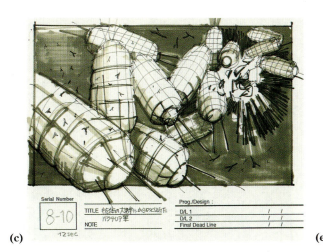

Credit/Debuchi (Programmer) Hi Tech Lab, Omnibus, Japan

(c)

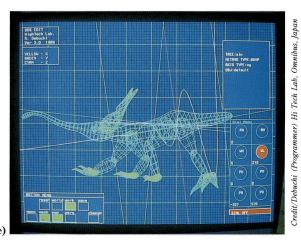

(e)

(d)

3 Small screen – even smaller budgets

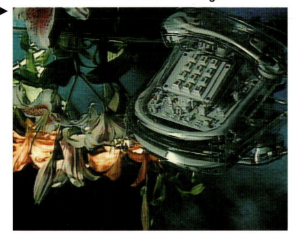 

Any mention of budgets can prompt designers to tell of stories of minimal or vast sums. Some managers and clients will say that graphic designers are rarely satisfied with the amount of money available for the production *work. Richard Higgs, a Senior Graphic Designer at HTV, describes, in Chapter 7, the resourcefulness required to design two very low budget titles: (above) HTV's* Helpline *and (below)* First Cut

Stills stores

Digital paint systems have integral stills stores but these do not have the capacity to hold much more than some of the work-in-progress. The concept of storing thousands of frames of full-colour images, with the ability to select and retrieve any one of them almost instantaneously, was crucial to the spread of graphic design as a service to broadcasting and fundamental to computer animation. The term 'frame store' is an alternative to stills store; this gives rise to 'frame grabbing' when a large number of frames are held and a single frame is said to be 'frozen' when it is found and recorded.

When the stills store principle is enlarged for television transmission to serve a news channel and other programmes a 'library system' is created.

Graphic designers are the involved in library management and the titling of pictures stored, so that the images can be searched by name via a keyboard. There is also a visual 'browse' facility displaying about 16 images on the monitor at once. Logica's 'Gallery 2000' and Quantel's 'Picturebox' are two such systems that have allowed graphics to be featured on-screen with a speed and ease not dreamt of only a few years ago.

Video artwork stored and filed electronically with instant recall in vast capacity stills stores like 'Picturebox' and 'Gallery 2000' now give graphic designers access to millions of stills

Graphic editing – the key tool in production

Trying to produce graphic information, still or animated, at extremely short notice – particularly for news – was until the late 1980s very frustrating for the editorial teams and for the graphic design departments. The electronic revolution emanated from newsrooms, first in the USA, as shown by this 'Almost instant graphics' feature on CBS News from the journal *U&lc*. (*U&lc* is wide-ranging graphic design review published by ITC, International Typeface Corporation, New York).

Television graphic designers working on news programmes at CBS in New York in the early 1980s told the mazagine U&lc that computers had now arrived to change the face of television graphics

The BBC's *Panorama*, or Thames's *This Week*, would regularly allow no more than a few hours overnight to plan and prepare artwork, leaving as little as two days for the production of one, or maybe two, 30-second film animations prior to transmission. This stretched the chain of storyboard, artwork, shooting, processing and film-editing to the limit. Obtaining statistics that had to be *final and unchangeable* many hours before the deadline was often an insuperable barrier to any co-operation.

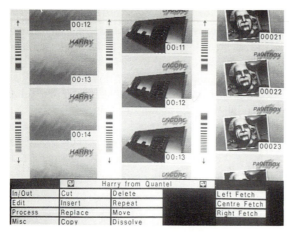

'Harry' edits images by 'cuts' and re-ordering but it can also replicate any of the selected video frames in seconds. Producing television graphic animation has been transformed dramatically by digital editing

Non-linear editing

Digital paint systems had brought image-making for stills into video but the computer had not, at that stage, found a way to compile sequences. The digital computer device from Quantel, strangely named 'Harry', did much to overcome this 'urgent news' problem with the introduction of 'non-linear' editing (the importance of this to graphic production is explained later) and then spread rapidly into every area of graphic work – programme effects, on-screen promotions, and thousands of commercials throughout the world. 'Harry' was first introduced by the makers at the International Television Symposium and Exhibition at Montreux in June 1985. At that time it offered only 80 seconds of random access capacity.

An extended version, with greater capacity and speed, unsurprisingly named '*Flash* Harry', joined the original marque. The storage capacity of 7,500 frames gives a maximum of five minutes of continuous video at 25 frames per second, which is more than sufficient for graphic design sequences.

Examples of the way the 'Harry' principle has affected graphic design abound throughout the world. It is rare to watch even a few seconds of title sequences, or commercials, without seeing it used in current transmission. All but one of the graphic productions gathered in Chapter 7 employed the 'Harry' device.

Within two to three years of work experience in post-production and facility houses its role in life has been grasped with enthusiasm by graphic designers and

technical staff. It has become as ubiquitous to television graphic design as the film rostrum of the first few decades.

News programmes in television are very powerful; they have large budgets and they were early in installing 'Harry' in their graphic departments.

A graphic designer's description of 'Flash Harry', prepared for this book by Paul Rigg, is given below. He is a Canadian designer, trained at Ravensbourne College of Design and Communication, who now works for Quantel.

Traditional VTR editing is a linear process where, to produce a result that is a combination of several sources of clips, a copy of all the material must be made, in real-time with each source clip being played at exactly the correct time. From this requirement mixer/switcher desks combined with edit controllers have evolved.

Editors working with film had no such problem. They could simply cut the film and stick in the new sequence in a totally random way, except that searching for the cut point meant playing the film as a linear process.

'Flash Harry' uses a totally random access store to hold the source information or video clips, and can recall any frame from this store in any order in real-time. This takes the idea of the film editor's random approach to editing even further as not only can the insertion of a new video clip be performed randomly but the location of the cut point is also random.

To make the most of this new-found freedom a new operating structure had to be devised. The traditional mixer and edit controller are not suitable as they require the video information to be presented in a linear fashion. As 'Flash Harry's' operation is closer to that of the film editor, why not mimic the mode of operation used by them? So the videola screen, now so closely associated with the power of Harry, was born. Combined with a pressure sensitive pen and a tablet to replace the film editor's scalpel and cutting desk, the new random access video editor was ready to go.

To many operators it seems as if the reels, that form the heart of the operating principle of Harry, also provide the power behind the tool. In fact it is the random access capabilities of the storage and processing that are the real power behind the tool, but without the flexibility obtained with the pen, tablet and the reels this power would remain trapped!

Therefore to say 'the power of Harry's reels' is not correct. It is the random access approach to video editing that makes Harry so powerful.

The diagram on page 62, prepared by Rigg (on a Quantel 'Graphic Paintbox'), explains the random access store and the non-linear editing process.

A younger relation known as 'Harriet', joined the family in 1989. Since the spring of 1992 there is also 'Henry'! The origination of the source clips that Rigg describes for manipulation on 'Harry' cover specially designed and prepared animation – from hand-drawn cel work, computer-aided video or film sequences, live-action or model shots made-to-order, library and archive film or video – or every possible combination of all of these.

'Harry' devices are consistently harnessed to the Quantel 'Paintbox' whereby whole runs of frames, from any such clips, may be 're-touched' frame-by-frame, extensively if desired, to alter or add to the images and change the animation. The 'Paintbox' link also enables rotoscoping and lettering to be generated using the same tablet and pressure-sensitive pen.

Too friendly?

A great deal of computer-based equipment is designated as 'user-friendly' to indicate that it is relatively easy to operate or that special computer programing ability is not required. This is true of both 'Flash Harry' and 'Harriet'. Due to this many of the most proficient operators are graphic-design-trained and their creative sensibilities have led the way. Once the interactive menu base has been mastered the inventiveness and the eye of the designer are essential factors. For some designers first-hand experience in operating any device gives them a far greater insight into its design possibilities, while they would not wish to become operators.

The virtues of 'Flash Harry' and 'Harriet' for any video sequence are described by Quantel as 'the ability to combine seamlessly all images, effects and graphics with limitless multi-layering and zero generation loss'.

Other systems

In discussing non-linear editing only the Quantel 'Harry' has been described for the sake of simplicity and, as can be seen from the examples of graphic work in Chapter 7 – 'The Reel World' – it is the 'Harry' system

that has been exploited by graphic designers at the time of writing.

There are now alternative systems in this market but comparing one electronic system with another is a minefield of options – high cost versus lower picture quality, high speed in operation against limited picture storage. And they are all being constantly updated!

News and current affairs programmes have again been the stimulus for a second wave of non-linear editing machines in this very competitive arena. In 1992 the BBC installed two new units called 'Avid' (Avid Technology Media Composer) – one as a post-in News and Current Affairs. Other non-linear offline editing systems are 'Lightworks', 'Eidos' and 'EMC2'. They too may become 'tools of the trade'.

Digital copying means that there is no loss of quality for each stage in the processing of an image. Anyone who has witnessed a photograph being copied from print to negative, then to print again, will appreciate this advantage very readily. (See note under 'Video signal formats', Chapter 6.)

Tools of the future

In section 8 of Chapter 9 the Creative Director Terry Hylton, who controls the application of one of the most comprehensive installations of electronic graphic hardware at the facility company SVC, responds to an enquiry about the acquisition of lower cost computer equipment like Apple Macintosh and quotes the recent production of a very effective 'storyboard/animatic' using the Macintosh by an independent design group.

Using Macintosh equipment and software to improve presentation methods has also progressed in the search for alternative tools elsewhere. English and Pockett were encouraged by one of their most recently recruited young designers, Jason Fisher-Jones from Middlesex University, to purchase a MacIIcx, a Hi-8 camera and other peripherals and programs (PhotoShop and Macromind). They used these to prepare the many graphic inserts for a pilot with simple animation, rather than static storyboards, a series about dancing, called *Hypnosis*, for Channel 4. The effect was so good that the Mac 'roughs' were edited for transmission. A 'sting' for an Anglia Television Christmas promotion was also broadcast from Mac output.

Enthusiasts, Darryl Pockett among them, have suggested that Apple should consider producing software for the potential market in the television design companies and in design schools.

Video stills

Obtaining permanent still images, known as 'hardcopy', from video monitors was a problem for some years. Direct off-screen photographs do not work.

The cost of the devices giving high-quality colour photographic prints or transparencies of video frames has dropped dramatically from the equipment of a decade ago. The early cameras of the Dunn Corporation and the Honeywell system were expensive by the standards of that time but they offered ways of eliminating the line structure of the raster in the video format and avoided the 'TV look'.

The hardcopy prints that graphic designers require to compose storyboards, to show programme staff and other colleagues their design ideas, to store 'work-in-progress', and for print publicity, are now much easier and more economic to produce. Images can be tranformed from all video formats and digital paint systems and they can be corrected for colour balance, density and contrast. Recently, video images can be transferred to Apple Mac and combined with high-resolution type and graphics, as required. They may then be output as 4000 line slides, prints, large display blow-ups, or introduced directly into print processing. Many of the illustrations for this book were prepared in this way by Print-Out Video Stills Limited.

Picture promotion

A further technical change that will affect all graphic designers is the prospect of widescreen transmission. The ratio of 3:4 has been the universal format but now 'PALplus' is being developed in a European collaborative venture. This new system might be in operation by 1995.

PALplus will present a screen proportion of 9:16 (a ratio of root3) produced by blanking-out bands at both the top and bottom of the existing television display on the current domestic CRT to give the 'Cinemascope' effect. The main problem in introducing change is to avoid designing a system that would only be available to a small section of the viewers and depend on the vast audience having to buy new equipment. PALplus overcomes this.

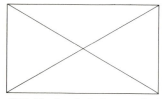

'PAL Plus' ratio is 9:16 and when transmitted on current screens 3:4 the picture is 9:14

Graphic designers will have to adapt to this markedly different ratio and it will be very interesting to see how they can make use of the format. Will the audiences really commit themselves to this new format? Will the broadcasters provide programmes in the required ratio? Or will the change be merely a passing fashion?

The optical advantages are not easy to prove but designers have simulated widescreen effects on occasions from the very earliest days.

The medium for everyone?

Prior to 1956 moving television images could not be recorded; this breakthrough had to occur before the micro-chip could be involved. Video tape recording and editing have passed through many stages, some very cumbersome.

Now the affordable video portapak, with instant playback, can be used to record almost anything, to be translated to prints to make storyboards or used as animatics. Pencil need never touch paper. The technical ease can be staggeringly powerful.

Marshall McLuhan, famous for his dictum, *The medium is the message*, went on to explain

> This is merely to say that the social consequences of any medium result from the new scale which is introduced by each extension of any new technology ... Gutenberg made everybody a reader. Xerox made everybody a publisher.

Would he have suggested that the affordable video camera makes everyone a television designer/film director and the desk-top computer generation of type makes everyone a typographer?

High-definition transmission?

Two HDTV video systems are in contention just now. They will have to broadcast at around 1250 lines to match 35 mm film quality and they promise far larger screens. (The European standard is 1250 lines and the USA/Japan 1125 lines.) Will there be several standards of high-definition throughout the world, continuing the present problems of conversion and equipment variation from one country to another? What effect will the changes have on designers and the facility houses who will work with them?

Summary

Any observer can see from the preceding review of the 'tools of the trade' that there is an overwhelming amount of equipment. Good relationships between designers and production staff are paramount. □

Chapter 7   The Reel World

Skilled model-making and lighting made this very natural effect for the title sequence of the BBC2 series called Chronicle. *Below the snorkel lens of the motion-rig that recorded the animation can be seen*

1 The Chronicle of *Chronicle*

Motion control animation for a BBC series

The opening seconds of this BBC title were so convincing as live action on first viewing that when the ruins were transformed into a model of the word 'Chronicle' there was a real surprise. Once again the element of illusion was well used by the designer and those who assisted her.

Liz Friedman's own chronicle of the production of this title is modest and succinct:

> This generic title sequence was made for a long-running title series about archaeology. My brief was simply to produce a design which would last for several years. My aim was to create an atmosphere appropriate to the subject matter and give a hint of the various historical sites likely to be covered.
>
> The movement started close to the model, panning across 'ruins' from many countries and periods. There is a Greek amphitheatre, Roman columns and an Egyptian pyramid, plus other structures. At first we seem to be in a real-life setting, then, as we pull out, we realise that each building is actually a letterform. Finally, we read *'Chronicle'* on a background of sand.
>
> The model was 14 feet long and the shoot was made with conventional 35 mm film camera mounted on a computer-controlled motion rig suspended from the studio ceiling with a periscopic lens. The Quantel 'Harry' was used simply to mix at the end from one take (lit with heavy shadows) to another take (lit more softly), to add legibility to the end 'logotype'.
>
> The music was composed to the final animation, not to the storyboard, by Richard Atree of the BBC Radiophonic Workshop.

Once her storyboard had been approved Liz commissioned line drawings from Peter Parr, an illustrator with architectural knowledge, to guide her chosen model-maker with details of Egyptian, Roman and Venetian edifices.

The decision to use three-dimensional models was based on the strong desire for reality and atmosphere. This could not have been achieved with even the most advanced 3D computer animation. Liz had worked with Doug Foster of Cell Animation on an earlier title, *War in Korea*, using models. She had early discussions with him this time before consulting a model-maker.

Alan Kemp is a freelance who has been making models for graphic designers since 1979. *Chronicle* took him nearly six months to complete. He contributed the idea of forming the letter 'N' from a broken pyramid. The objective was to represent the letters with as little distortion to the real items as possible. Alan's voyage into graphics is told in the first section of Chapter 8.

Motion control

The effectiveness of making the model appear genuine in those first few frames relied very much on the camera operator and the lighting. Once the model was ready Liz returned to Doug Foster and Cell Animation's computer-controlled motion-rig. This gives motion blur,

to help smooth out the moves, and Foster uses film exclusively on the motion-rig because less light is required on the models. The resulting smaller aperture gives a much greater depth of field, making models look bigger and enhancing the scale.

A Telecine transfer was made to digital D1 tape when the editing was required on Cell's in-house 'Harry'. Foster's route from designer to camera work is given in Chapter 5 and the value of the modern motion-rig in graphic design for television is given in Chapter 8.

Golden sands

Liz Friedman's hope that her design would be used over a long period has been enhanced by three accolades. The Broadcast Designers' Association gave a Gold Award in Las Vegas in 1990, the Royal Television Society gave it a nomination in the same year and in 1991 it gained a Gold Award at 'Imagina' in Monte Carlo.

Credits for Chronicle
Graphic designer: Liz Friedman
Model-maker: Alan Kemp
Lighting and motion-rig camera: Doug Foster/Cell Animation

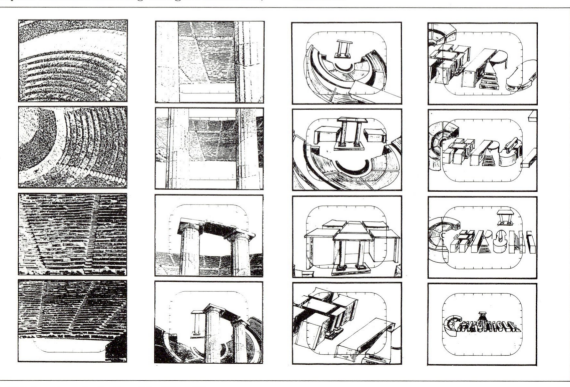

Liz Friedman's early storyboard for Chronicle

2 Dissecting a medical series

Computer animation at NHK/Tokyo

The Universe Within is a medical science series of six programmes made by the Tokyo-based television company NHK (Nippon Houso Kyokai/Japanese Broadcasting Organisation). This was first broadcast in Japan and several other countries in 1989. A series is now in production for transmission in 1993.

The makers said they would illustrate 'the functions and the phenomena deep inside our bodies which are unknown and invisible – even under the most powerful microscopes'.

Applying computer animation

Computer-aided three-dimensional animation has given graphic designers the prerogative to construct, light and move within imaginary spaces on a scale far beyond the capacity of the most complex hand-drawn animation. This command is particularly well applied to scientific information. The army of artists required and the immense difficulty of controlling the complex detail on every cel in film animation, to achieve a comparable result, would be impractical.

The Japanese took to computer-animated graphics with their predictable intensity. In this they have already contributed so much to this area of graphic design that it is no surprise that electronic media has been utilised to great advantage by NHK in this project. The graphic design department was briefed to produce a massive 30 minutes of graphic animation for the total of six one-hour-long episodes on the human body.

The themes were: sex and reproduction, muscle and bone, digestion and absorption, the liver, the heart and circulation, and the immune system.

The series' art director Kasuo Sasaki describes his department's contribution to this vast undertaking:

Most projects are relatively simple. Designers at NHK expect to draw a storyboard, based on a brief, and then hope to gain approval for their plan fairly quickly. Once gained, the next stage is a discussion with the 'production people' to agree the methods required for the animation. A schedule and budget based on these decisions will then be fixed. Throughout the making process the graphic designers are responsible for delivery on time, the final cost, and the quality of the completed work – precisely the way they work all over the world.

Everything was far more complicated for us in a large scale project like *'The Universe Within'*. The very long production time, the high output with many sections which had to be visualised with a common concept and style, and an extremely high budget to manage meant we had to ensure the system was efficient and stable. We could not work piecemeal. Control would be lost and the costs would spiral.

One significant thing we did was to arrange for a computer-graphics production room to be dedicated absolutely to this project. At the peak of the production we had three Iris workstations and other devices including: 'Links', TDI, 'Symbolics' with a paint system, a 'Simaseiki' paint station and several VTR facilities. We hired three enthusiastic animation artists to operate the room. (They were university students who worked for us part-time.) Two fellow designers worked for about half the production period while I worked full-time. Three programmers were engaged and they were students and freelance staff.

Our role did not stop at designing and drawing the storyboards but entailed constantly managing the facilities and educating the non-experienced staff. It was a demanding but challenging task and we were able to deliver about 15 minutes of fully three-dimensional animated sequences on time.

The outside contribution

Another 15 minutes of similar output had to be commissioned from two production houses, outside NHK, and we had to establish a special relationship with them. The unusually large amount of high-quality work was certain to involve them for almost a whole year. They were: the High Tech Lab Company who involved two technicians, two in-house directors, five designers as artists and operators of 'Wavefront'; the other company was Omnibus Japan and they used two programmers and three designer/artists, one to operate 'Prism'.

We guaranteed that the project would last for the fixed span and for that they would co-operate and support us as though they were an internal part of NHK for that period by exchanging technical information working at a special rate and below the usual cost for each second of graphics. These costs cannot be disclosed. High Tech Lab Japan and Omnibus Japan have excellent reputations here and internationally. They have made dramatic presentations to the SIGGRAPH video reviews for many years.

$G=(\sum F_i^2)^{1/2}$, we cannot yet determine the substantive meaning of the derived scales. It seems at present, we can only theoretically formulate the summated scales with some simplified all-1 and 0-or-1 weighting schemes, including those represented by the above hypotheses. When the all-1 weighting scheme is applied, however, the validity of a scale based on components instead of individual items may not be improved, since they are "too precise" (in terms of the clarification of dimensionality) on one hand and too rough (in terms of the uncertainty of the relationships among factors being pooled together) on the other.

The rule-of-thumb represented by the conventional "expert" approach should not be the only way for better weighting, however. Compared with the case of engineering science, we have more expedient statistical means for factoring in psychosocial studies. Therefore, efforts ought to be invested in future studies to work out feasible strategies with more solid theoretical and empirical grounds. There are many ways of constructing a scale that can be explored, indeed. For example, we may try to take the absolute values of the components before summing them up. There must be more organic and accurate ways for integrating the results of a dimensional analysis. Besides, there are some empirical means available, which can be used as an alternative or supplement to the theoretical approach explained in the above.

Weighting and multiple regression: A practical approach

The empirical or practical approach to unidimensionalization can be formed by summarizing behavioral and social science research experience in weighting different items of a scale. Notwithstanding the fact that we had to resort to scientific/engineering ideas and methods, psychosociometricians are not to blame for totally failing to provide research workers and practitioners with a way out. Technically they have derived numerous statistical means and passageways based on which a practical approach to the issue of ultimate unidimensionalization could be figured out. Indeed, behavioral and social scientists have the advantage of including samples of study objects (subjects) large enough in most single studies that make such an approach not only feasible but also expedient. A knowledge of unidimensional scaling is fundamental in understanding this. One important task is to determine the relative importance of component items or subscales, that is, the weights they carry in the composite scale.

There are two kinds of weights. The nominal weights are those we deliberately

The storyboards I had to prepare for *The Universe Within* and the procedure was different from the norm. Because the work was deeply dependent on intricate medical knowledge, we needed expertise for each of the subjects covered. There were many meetings with doctors and medical specialists, as well as the television directors and other designers.

A vital function of the storyboards was to obtain clear agreement amongst people of different professions and ways of thinking, on the appearance of the images and the storyline. I drew the boards over and over again until everyone was satisfied.

Counting the cost

Production is never commenced until we get the sanction for our prediction of the budget from the producers. With experience this is normally not too difficult but the scale of *The Universe Within* made it really hard. The sequences for each section had to be created simultaneously and in relation to each other. To keep to the budget and to control the quality of the output step-by-step meant talking with those in charge at every stage.

Most UK television stations seem to operate a 'total cost system'. NHK do not at present. Therefore we do not calculate the cost of staff time or the facilities inside NHK as we have to for outside resources. This means we may use whatever is available, from studio work to post-production and computer graphics. The programme producers have only to 'book' what they need. Unfortunately (and I regret to say this) the level of skill and the facilities are not always as high as those outside – I have heard the same complaint from graphic designers in the UK. At present about 50 per cent of computer-aided animation and post-production for NHK is commissioned.

NHK's graphic organisation

There are two main divisions for graphic design. One is called Art Division and the other is Design and Operation Department of NHK Art Company. They have a complicated history and in common with other public broadcasting stations in the world we now have controversies over this topic – as for instance the BBC.

I belong to the former division and in one way it is, I believe, unique. We have designers employed to carry out work in graphic as well as scenic design on a single television programme. For example, I did the set design for the studio presentations required for *The Universe Within* whilst art directing all the graphic material. There is dissent over this. Some feel it is not the right way as the separate highly individual skills in each side of the profession will be lost. For my own part it was good to work as the 'total concept designer' for the programme – not just on one aspect.

Our role as designers is to plan as globally as possible. We are not concerned with the hire of operators of equipment or engineering facilities. In this we are, I think, very similar to groups like English and Pockett and McCallum Kennedy D'Auria in London.

The NHK Art Company has a greater spread of skills than we have and it employs graphic designers, digital paint system operators, character generator operators and photo-setters.

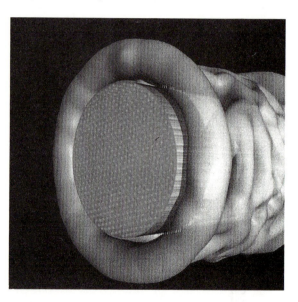

(Opposite) some of the hundreds of storyboards designed and drawn by Kazuo Sasaki, of NHK, for the animations for The Universe Within. *Many were changed several times to gain the approval of editorial staff and medical advisors. (Above) Preliminary computer output to show the appearance before rendering multiple frames*

Credits for The Universe Within
Executive Producer: Katsuhiko Hayashi
Producer: Hirofumi Ito
Director: Masakatsu Takao
Designer: Kazuo Sasaki

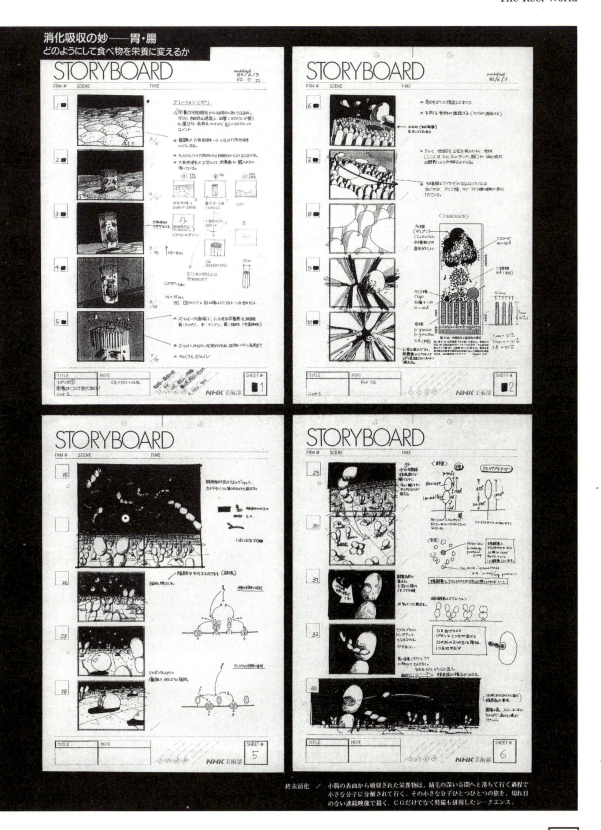

3 Small screen – even smaller budget

Programme titles produced on a shoestring

In attempting to describe 'the real world' it is only fair to include a description of graphic work where there is very little money. Richard Higgs is one of the few graphic designers to have worked for ITV and the BBC. His account here explains that in much television design the onus is on the designer to be very resourceful with little expenditure.

What is a budget? Let me first say that talking to designers about budgets is like talking to fishermen in a pub. Whenever you mention them, apocryphal tales of the biggest overspend of all time, or how they achieved a complicated *Star Wars* effect for 10 quid, will be boasted. Within the television companies it is only in the last few years that designers have had to account for every penny they spend, including what used to be hidden costs, such as their time and internal resources. A budget now includes the designer's time, the equipment used (edit suites, studios, cameras, dubbing, etc) and any real cash costs incurred.

What is a small budget? The average cost of a network title sequence and programme styling will be between £15,000 and £35,000 and a small budget could be from nothing to about £5,000.

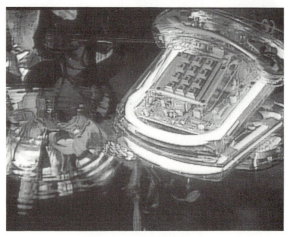

Richard Higgs's design secret was the low-budget for the HTV Helpline *animation was achieved partly by using a 'prop' that created a lighting effect*

Money management

External costs are usually items where the price is negotiable. You can go to an animator, model-maker or scenic painter, telling them how much you have to spend – playing the 'wheeler-dealer'.

Internal costs are fixed hourly charges made for 'Paintbox', edit suites, character generators, and other services. One difference between the internal and external costs is that external costs (real cash) leave the building, whereas internal costs can be 'massaged' on time sheets and shared between other programmes.

The duration of the title sequence can be determined by the music. Obviously the longer the titles the more they cost. Often the graphic designer will be involved in commissioning and choosing the music as it is a vital part of the finished sequence.

No designer is ever really sure where the ideas for title sequences come from. Usually after exhaustive research and many frustrating hours over a layout pad, an idea will strike and immediately become the obvious solution. References, props and other material relevant to the subject are gathered over this period.

It is important to have a thorough understanding of the equipment and resources that are available to you, so, when an idea emerges, you know that the result is achievable within a given budget. Often, on a small budget, it is useful to find objects that move (e.g. a clockwork rabbit or a tank of live goldfish) and provide moving images that are cheaper than animation.

Wheeling and dealing

Two examples of low budget work: first, a sequence for a series called *First Cut* where local bands, who had not previously made a record, could compete for the prize of a recording contract; and second, a four-second opening title for the regional public service announcements, *Helpline*.

The budget for *First Cut* was about £2,000. 'Codsteaks' *(see Chapter 8, section 1)* built me a silver proscenium with working curtains, about three feet wide. I spent a day with the cameraman, Doug Hartington, shooting on Super8 various images with 'cut' as the central theme. We shot cut flowers falling in slow motion, we cut a birthday cake with one candle, breaking billiard balls, a deck of cards, a spinning Swiss army knife and a variety of strange abstract images. When the model was completed we set it up in a small room and projected the films on to the theatre with slides and pattern lamps and smoke. Then Doug and I re-shot close-ups of the proscenium arch on BetaSP and then edited them to a short piece of re-mixed soundtrack of drumming.

A break-down of the costs: the model was £700; a day and a half with the cameraman, including the hire of the Super8 was £160; a day in the edit suite was £600; the props – flowers, cakes and goldfish – were £40, and my design thinking time was £500.

I was told that the budget for *Helpline* was nothing or as close to that as I could get! I worked with Peter Ming Wong who suggested we use a neon-illuminated telephone. This was hired for £9. We had a large mirror made, with a shallow rim, which allowed us to cover it with 5 mm of water. I painted a piece of polyboard with emulsion paints and spray cans. A bunch of flowers cost £2.50. This was set up in our news studio, on the news desk, and shot using a studio camera in the few hours between bulletins. The soundtrack was put together using two chords on the in-house audiofile; the logo was created on our digital paint system in a couple of hours. Finally the pictures were slow-motioned and the graphics added in an hour's edit.

We had spent £40 on the mirror and £11 on props. The other resources and our time was 'lost' within the routine programme budget.

Producers get what they pay for, and the larger budget allows the designer the opportunity to be more creative.

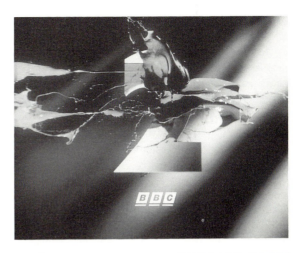

4 Channel identity/BBC2

An independent design group commissioned by the BBC

Trademarks and company symbol design have always been the most contentious and difficult of problems. The stakes are high: the rewards can be high. The constant and prominent use of a design identity, over periods of many years, by television stations is of vital importance to them.

If politics is 'the art of the possible', as R. A. Butler once wrote, Martin Lambie-Nairn is a practitioner par-excellence. He has always made things happen. When talking to him the concept of a graphic designer as a lone figure at a drawing-board vanishes instantly.

For his company (previously Robinson Lambie-Nairn, now Lambie-Nairn & Company) and for the profession, he epitomises the designer as entrepreneur.

At one point Martin reminded himself that he started out as someone who was primarily a craftsman who respected hand skills above all else. He shied away from the word 'politics' to avoid the feeling of manipulating people rather than advising them.

In the recent creation of a revised Channel identity for both BBC1 and BBC2 his foresight and powers of persuasion can be seen to be of the highest order. His company had previously produced corporate design projects for Anglia Television, TF1 in France, and a decade ago the highly adaptable Channel 4 material.

Redesigning the on-screen identity for both BBC channels was the largest television graphic design commission to be placed by the corporation – a long way from the drawing board and sharpening pencils.

Reasons for change

Tensions of competition and talk of retaining audience share led the BBC to commission a new corporate identity for their radio and television channels' print and display work. Michael Peters PLC carried out this task and the new logotypes were soon applied to stationery, vehicles and in the *Radio Times*. Those for BBC1 and BBC2 Television were resolved without any consideration of their adaptation or suitability for animated on-screen use.

'Design evolves by gaining the confidence of your client', said Pam Masters, Head of Presentation at Channel 4, 10 years ago when she was responsible for nurturing C4's proudly famous symbol designed by Martin's company. She came to the BBC in a similar capacity when Lambie-Nairn & Company were hired to produce a report, at the direct request of Sir Paul Fox, to advise the BBC on the way all television presentation should be directed in the environment of fiercer competition. 'As one of the world's foremost programme-makers it is important that the BBC takes a similar lead in establishing distinctive identities for its two television channels', Pam Masters pleaded.

In preparing his report Martin remembered the warring factions that had occurred elsewhere, certainly between the graphic designers and the presentation managers at Channel 4, and he took the view that unless you solve that kind of problem you will never solve any other: 'We had previously recommended a change of structure at Channel 4 and reorganised the graphic design department by employing younger designers. That changed the whole climate of design work.'

Lambie-Nairn decided to use audience research to find out what people really thought of the BBC television presentation, rather than make guesses – inspired or not. The results stressed the pre-eminence of keeping the two channels, clearly perceived by the viewers, as quite distinct while retaining the past strengths of the corporation. The computer-generated version of the revolving globe, which was now to be reconsidered, had been introduced in February 1985 and the BBC 'TWO' animation at Easter 1986.

The continued use of the audience research unit within the BBC was also recommended to guide and validate all future presentation design. 'Feedback' was the coded message.

Selection of those who would carry out the work was to be by tender from a number of groups, who would be paid to make their proposals. At this stage Martin's company stood clear, but they were invited to be one of three contenders to answer the brief. English and Pockett and a section of the BBC Graphic Design Department were the other two.

Lambie-Nairn and his team were chosen to carry out the scheme they had submitted.

Designs were prepared and produced for on-screen animations for both channels. For space reasons only the results of BBC2 are discussed here.

Themes and variations

The brief asked that the idents should be extended to a number of versions to overcome boredom through constant repetition and to allow the presentation department some choice in which version to use, relating them to the mood and subject matter of programmes whenever suitable.

Martin's co-designer was Daniel Barber. He had joined LN&C directly from Central/St Martin's College of Art just three years before. They did not present storyboards with frame-by-frame precision to relay their ideas but simple 'concept sheets' to show the composition, the themes, the colours and the chiaroscuro they envisaged – much as fashion designers do.

Each variation was to be played in the same key. The results have an elegance and wit, giving them what the animator Norman McLaren described as essential elements for a work of art, 'Cohesiveness, but not so much cohesiveness as to become dull – there must be surprises!'

Daniel has been pleased that the alternatives have been used subtly to suit the mood of the surrounding programmes. The nine 'symphonies' composed to date are: moving water, wind-blown silk, splashed paint, paper, copper – with arcing electric sparks, a guillotine-like steel blade, glass, a neon-lit figure, and a version called 'shadow'. He was cautious about extending the ambiance too far and rejected the more bizarre games played, for example, by MTV, the music channel, with their on-screen idents.

Alan Yentob, then Controller of BBC2, welcomed the changes, saying 'it has more impact and flexibility than our current on-screen image'.

An important but backroom aspect of the project was the need to consider the national and regional identities. This led Martin and his team to a close involvement with the BBC Engineering Department to rationalise the vital transmission arrangements throughout Scotland, Wales, and Northern Ireland and England.

Further versions expanding the theme of these BBC2 idents have now been prepared by the BBC Graphic Design Department.

Filming the sequences

All nine of the original versions were shot on film at Cell Animation. The camera was *locked-off*, that is in a fixed point in each case. Motion-control was not required. They were shot at the rate of one sequence per day. Tests were made to check various points, and the splashed paint version was the most complex, with the test taking half a day. How would the paint fall and how should it be released? Finally it was decided to film at 350 frames a second. The moving water version was shot at high speed too, but at only 100 frames per second. The sparks on the copper panel were real, made by an arc-welder, not added electrically, but they were slowed down in post-production. Under the billowing silk was the perspex figure '2', about 24 inches high, on a large sheet of glass because the first plan was to underlight the sequence. In the end it was top-lit only. The area underneath the very rigid scaffold erected to ensure there was not the slightest camera movement was about 8 feet by 12 feet. At this size there was no alternative but to seam the silk. One complete piece would have been best.

Credits for BBC2 on-screen identity

Designer/Directors: Martin Lambie-Nairn and Daniel Barber
Producer: Celia Chapman
Lighting/camera operator: Doug Foster at Cell Animation
Model-makers: Steve Wilshire Creative Effects (Water/Silk/Paint/Paper/Copper)
Asylum Models and Effects: Bob Hinks and Mark Curtis (Blade/Glass/Sign/Shadow)
'Harry', 'Paintbox', and 'Encore': Rob Harvey Cell Animation
Telecine, Mastergrade, and Digital Editing: SVC Television
Music: Anthony and Gaynor Sadler at Logorythm Music

5 'All that hoo-ha at the beginning'

Documentary titles reach the headlines

Television

Stalin's fiery story dampened by talk

LAST NIGHT'S VIEW
By PETER TORY

beginning of one of those loudly-heralded, long-prepared and extremely costly series about momentous matters of history.

In this case it comes from Thames and is the story, told in three weekly parts, of Joseph Stalin.

Stalin is all worthy stuff. Indeed, tremendous credit must go to Jonathan Lewis who wrote, produced and directed it.

However, the main problem with such programmes is that there are endless and very ancient talking heads — in this case most of them speaking in Russian behind flickering sub-titles — and one often feels that one might almost be as well off and certainly more comfortable sitting in bed reading a book about the subject with suitable music turned up on the gramophone.

No, for me the most exciting part of these documentaries is all that hoo-ha at the beginning.

Oh, to be in charge of all those symbolic anvils, scorched maps, blood-drenched flags, boiling clouds and crimson fire-storms on the horizon.

A huge hammer smashes down on red-hot metal bringing an explosion of sparks whilst a farm-worker's sickle slashes across a burning sky . . . a heavy style of orchestral music thunders about our modest sitting-room frightening the cats . . . On a great anvil, spelt out in smouldering steel, appears the word STALIN. Flames and smoke swirl. Crash, wallop, boom . . . for me the most exciting part of these documentaries is all that hoo-ha at the beginning. Oh, to be in charge of all those symbolic anvils, scorched maps, blood-drenched flags, boiling clouds and crimson fire-storms on the horizon.

This was Peter Tory's review in the *Daily Mail* on the first episode on the Thames Television's documentary series on Joseph Stalin. Morgan Sendall, who is now a freelance graphic designer in London, describes his work in designing the 43 seconds of these titles and his amusement at such exuberant reactions.

It is a rare pleasure for designers to be singled out and praised by the national press, tabloid or otherwise. I might add that Mr Tory is not a personal friend, and I did not pay him to write the article!

Abundance of imagery

The title sequence for Stalin was a godsend in terms of source material. There is an abundance of graphic revolutionary and propaganda imagery, much Stalin 'iconography', as well as some extraordinary archive film. I decided to use the best of this material I could find, but nevertheless a strong central theme was needed that progressively focused on Stalin. The Bolshevik symbol of the blacksmith's hammer, and the fact that Stalin's name means 'steel', sparked off the idea of having the Cyrillic letters of his name hammered out one by one. I felt it was a suitable metaphor for Stalin's ruthless rise to power, blow by blow.

The sequence was initially shot on film at a genuine blacksmith's workshop. The letters were mostly pre-cast in metal and heated until they were red-hot, then hammered on an anvil in various ways. The final explosion was created by wiring a small charge under the hammer. The Cyrillic letters were carefully replaced by Roman ones so that a smooth transition could take place beneath the smoke.

Multiple techniques

Another Bolshevik symbol, the sickle, provided a parallel theme of aggression by using it to hack murderously through images of Stalin's victims. The effect was achieved by scything apart tracing paper back-lit with key colour. The images were then keyed-in later on 'Harry'.

The opening track down a corridor was shot at Camden Town Hall. All other live-action shots were originated from archive, most of which needed re-processing through 'Harry' in various ways. The flag-waving heroic figures and the surreal scenes of a giant Stalin destroying the masses were all hand-animated. The latter were inspired by a little-known Soviet satirical painter called Pyotr Belov, whose work one of our researchers stumbled across in the Soviet Union, as it was then. All the sequences were put together on 'Harry' to the required degree of refinement.

Finally, a strong, Russian-sounding score by Nigel Hess brought the whole piece to life (and frightened the cats at the same time).

Credits for Stalin

Camera: Doug Adamson and Jim Howlett
Model-maker: Steve Wilshire
Animators: Boxer Animations
'Harry' operation: Michael Parry
Music: Nigel Hess

6 Designing for two worlds

A BBC co-production

The enigmatic title *283 Useful Ideas from Japan* came from a book. Morgan Almeida, a graphic designer at the BBC, describes what went into his mere 15 seconds of screen time designed for a co-production between the BBC and NHK of Tokyo television. This gained almost 283 awards, among them a bronze at the Broadcast Designers' Association of America in LA, exhibition at 'Imagina 1991' in Monte Carlo and at the Film Festival in Geneva.

> The programme's intentions in these 13 10-minute episodes was to show the extreme diversity of practical, innovative and often weird ideas which emanate from Japan showing their culture, their obsessions and their perfectly packaged lifestyles.
>
> Each episode covered items which ranged from the human bonsai and pornographic love hotels to women's wrestling and Audrey Hepburn worship, Japanese advertising to baby consumerism, computer graphics to production-line, pop stars to toilet technology. Essentially a modern 'Japanorama' – a word once considered for the title.

Concepts and contents

The intentions behind my title design were to avoid all those brushstroke calligraphic clichés and patronising historical references to samurai and geisha girls, and to make progressive interpretations of contemporary Japan all to be contained in a package

exported from the East. The concept hinged on the idea of the perfect 'Japakage' bursting with ideas, constantly evolving and mutating in a swirling vortex.

Every stage of the animation refers specifically to each of the 13 episodes: out of the exploding atomic inferno comes the phoenix-like rise of New Japan, revealing chaotic Tokyo, followed by women's wrestling, the human bonsai, then chattering Coke cans in the world of Japanese advertising, love hotels, pachinko (pinball machines) and tattoos mutating into a Tokyo toilet which is in turn gobbled up by 'Gastronomica Japonica' (one of the episode titles) a flash of computer wireframe and a gigantic red ball symbolising a modern-day Japanese flag, the idol worship of Audrey Hepburn and television manipulation – all in front of relevant video tape footage.

The frenetic box finally resolves into a minimalist logo that deliberately combined the Nipponese symbol 日本 with the English logotype (a

feature only the Japanese co-producers would appreciate). Then the large red Japanese ball crashes home into the white screen.

Only 15 seconds long due to the short duration of the programme – blink and you'll miss it.

How it was achieved

Stop-frame motion animation was shot against Chromakey on a rotating, elevating rig, then keyed against digitally manipulated video footage to create the swirling panorama. This was combined on 'Harry', with digital editing and computer graphics.

The title was completed for an extremely limited budget of £9,000; content graphics and customised end credits were then added.

Credits for 283 Useful Ideas from Japan
Graphic Design: Morgan Almeida
Post-production: Rushes
Model mechanics: Tony Palmer
Music: Anne Dudley

7 Method into *Madness*
A documentary series for the BBC

McCallum Kennedy Auria's portfolio of work completed since they started at the beginning of 1989 is bursting with prominent series titles for the BBC, a healthy mix of ITV companies, and worldwide commissions for commercial and corporate identity – the ideal now confidently anticipated by the television design groups formed in Britain in the past decade.

The titles for the series *Madness*, presented on BBC2 by Jonathan Miller, were commissioned by an independent company, Brook Productions, and the art director was Graham McCallum. The simple blue monochrome gave a cohesiveness to the writhing and tormented movement of the main figure. Graham's starting point came about by looking at Victorian phrenological heads:

A long understanding of the qualities of film were continued in this title for Madness *but even here electronic 'seamlessness' was brought in to gentle activity via motion-control and 'Harry'*

> These ceramic heads, with their craniums divided into compartments by black lines have been noticed many times before for their graphic potential.
>
> The idea was taken a stage further by marking a real human head with words and symbols relevant to the subject matter of the series. Half a day was spent in make-up applying rubdowns and writing with felt pens directly on to the head of a suitably bald head.
>
> This long-suffering person then spent the afternoon lying on his back with his head clamped into position. This allowed for the moves to be plotted on a motion control camera.
>
> Archive footage, found during the research stage of the production and cut into a sequence, was then projected on to the head for the final filming. The projector also provided the light source.
>
> The sequence was finally cut together and the colour re-graded in 'Harry'.

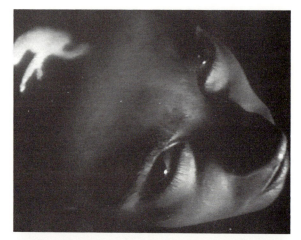

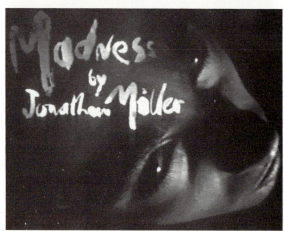

Credits for Madness
Director/Designer: Graham McCallum
Producer: Jayne Marshall
Client: Richard Denton – Brook Productions
Motion control camera: The Mill
'Harry' operator: Tim Burke at Cal
Music: Duncan Browne

4   Channel identity for BBC2

On-screen television channel idents were for too long unvarying symbols with repetitive jingles. Lambie-Nairn and Company have confirmed the move away from rigidity with a cluster of nine signals to engage the interest of viewers of BBC2 programmes. The stills printed here are: (a) the surprise of a steel blade-like '2' falling with a percussive sound track, (b) a gentle passage of wave motion at an eccentric angle,

(c) sudden electric sparks arcing to echoing clangs, (d) the shock of splashed paint and (e) wind-blown silk to reveal and conceal the figure '2'. None of the sequences was longer than nine seconds. The politics of commissioning and designing high-profile and long-term representations of a station's corporate image are considered in discussions with designers in Chapter 7

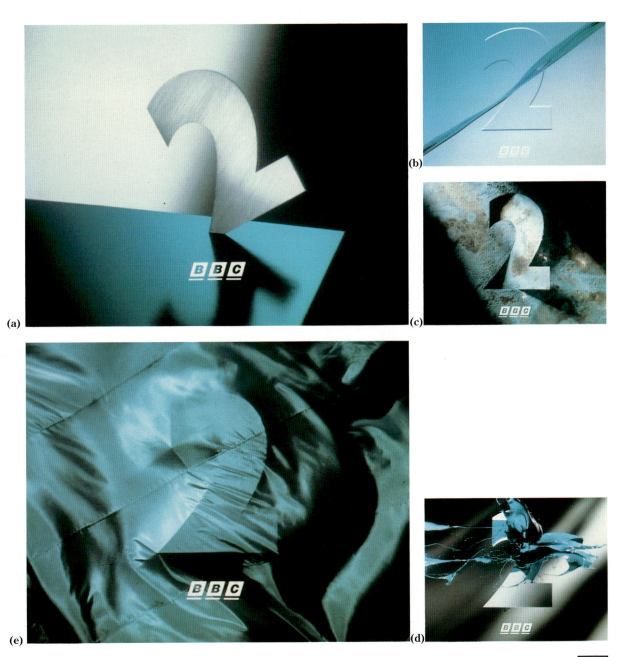

(a)

(b)

(c)

(e)

(d)

assign to components; the effective weights are those the components actually carry in the composite (Ghiselli, 1964). The two may or may not be the same. The point is, "In order for the effective weights to correspond to the nominal weights, we have to be sure that the components carry equal weight to begin with" (ibid., p.298). The means by which to achieve this is a procedure of transmuting the scores on the components to standard-score form to equalize the weights they originally carry. When this is done, we can establish the relative role of individual components exactly by differential weighting according to our theoretical understanding and/or based on some empirical grounds. Taking a look at psychosocial research literature, however, one can get surprised that few, if any, studies have emphasized weighting in the development of needed scales. And this situation appears to have been justified in practical terms. The main grounds can be summarized as follows: a) The problem of assigning weights to items in a scale is one which is rather annoying (Sewell, 1941); b) It is difficult to defend other arbitrary systems of weighting in comparison to an equal-weighting scheme (Nunnally, 1978); c) The problem of assigning weights to items in a scale is not of great practical significance in light of the roughness of most sociometric devices at certain times (Sewell, 1941); and d) Studies have shown that essentially the same final results are obtained with arbitrary common sense weighting as with more complicated, but still arbitrary, statistical techniques (ibid.).

These reasons, however, do not warrant desirable results and are no longer tenable in face of the demand for more accurate measurement nowadays. For example, McIver and Carmines (1981) indicate that the interpretation of Likert-Scale scores falling between the extremes is problematic except in relative terms. Indeed, the results obtained in such a manner cannot fully meet the needs of practice, which often requires a cutoff point as a criterion for diagnosis, classification, and differential treatment. As a matter of fact, the function of differential weighting is "to make the composite more precise or reliable, more meaningful, or more predictive of some other variable" (Ghiselli, 1964, p.293). The question, thus, is not whether or not differential weighting, where applicable, is desirable but whether or not it should be done arbitrarily. According to the new norms set in the preceding pages, a theoretical/empirical basis has become necessary to maintain a scale as a more precise and meaningful composite mea-sure.

If unidimensionalization is done mainly through combining different factors or aspects rather than simply dropping some of them, the issue of multidimen-

5 *Stalin* – an ITV documentary

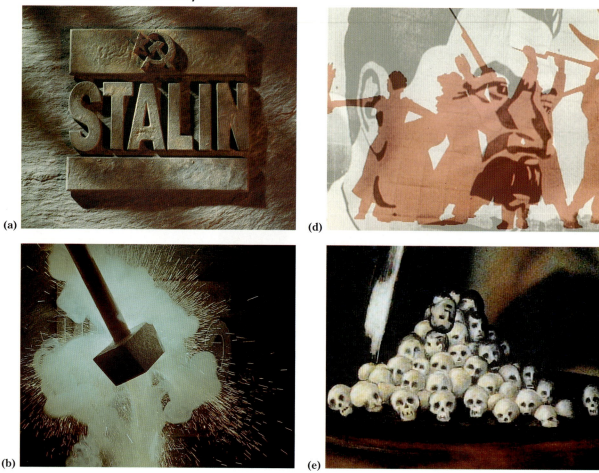

(a)

(b)

(d)

(e)

(c)

Morgan Sendall's series titles, made when he was a Senior Designer at Thames Television, were greeted with quite ecstatic praise from the television critic of a national newspaper. The quotations from the review which amused Sendall so much appear in Chapter 7. The 43- second animation exemplifies the current production mode of welding together diverse techniques. In Stalin the mix is live-action filming (**b**), using three-dimensional models (**a**), clips of black-and-white archive film (**c**), and artwork and hand-drawn cel animation (**d & e**), – the surreal scenes were based on the work of the Soviet artist Pyotrv Belov whose work was discovered by a programme researcher. The designer, now an independent consultant, says: 'All the sequences were put together on "Harry", confirming the very widespread use of this graphic design tool'

6 *283 Useful Ideas from Japan* – an in-house BBC design

During the production of this 13-part series the title 'Japanorama' was considered because the programme's content was so wide ranging. The stills on the opposite page enhance the account, given in Chapter 7, of the explosive way the BBC Graphic Designer, Morgan Almeida, managed to 'package' some of the exotic subjects in this opening sequence

7 *Madness* – for the BBC by an independent group

McCallum Kennedy D'Auria is one of the recently formed graphic design companies set up outside any broadcasting organisation and now able to work for anyone, anywhere. The pattern of MKD's formation and their current work is contained in Chapter 2, 'The Framework of Television Graphics' and in Chapter 7 Graham McCallum explains the design process.

This opening title sequence for the series Madness was produced using a mixture of archive film, obtained during the making of the series, and film shot in a motion-rig studio session

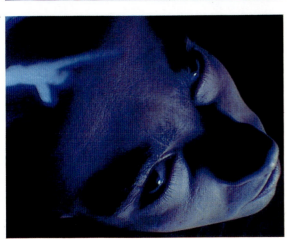

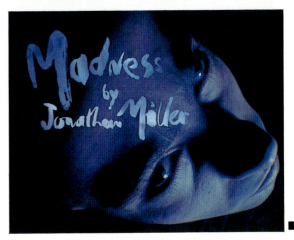

8 *The Heart of the Matter* – a design for the BBC via ITV

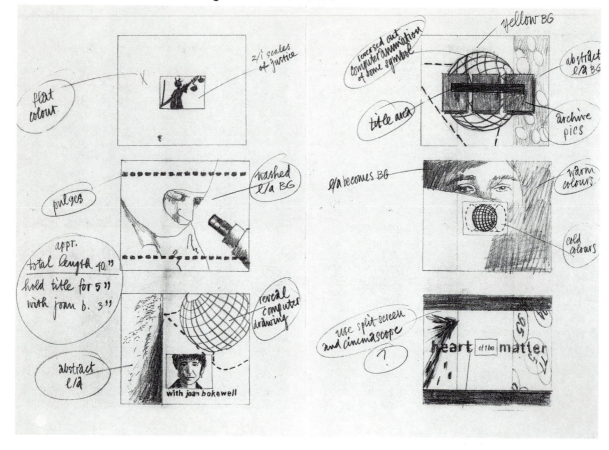

The loosely presented storyboard prepared for this series title by Chris Jennings and Jean Penders (above) was backed up by coloured 'stills', like the two samples below, made on a Quantel Paintbox. These gave the editorial team at Roger Bolton Productions enough details to discuss design refinements and approve the production process. The circumstances of this commission are given in Chapter 7 and final stills are shown opposite

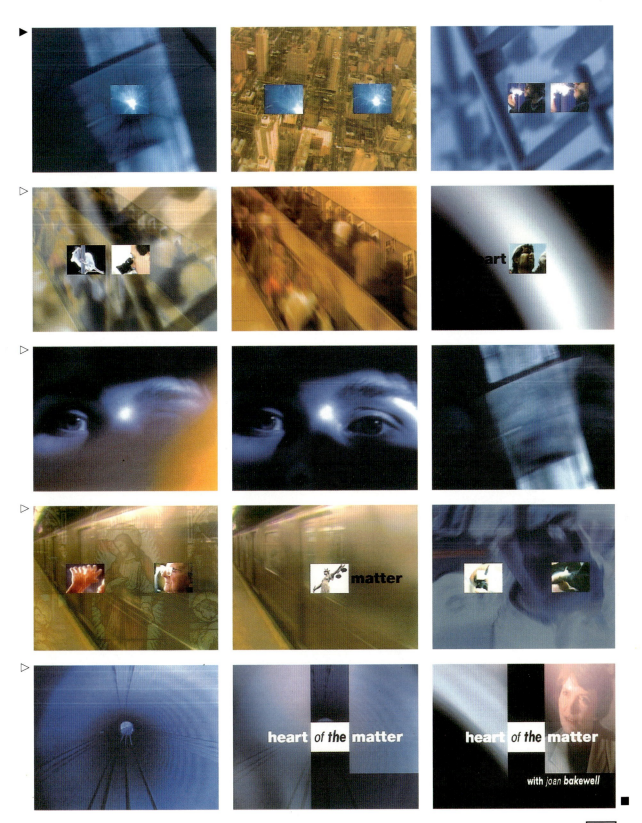

9 _MC Classic_ – a commercial made with a facility house

(a)

(b)

Art Director Barry O'Riordan of The Moving Picture Company describes in Chapter 7 the complex production of this television commercial. These stills from the final animation show: (a) chocolate-coloured satin envelops a honeycomb-covered spinning double bass – shot in reverse; (b) and (c) show the amalgamation of the live-action dancers and the main character with the computer-generated 'wrapper' hands, arms and body

(c)

10 *2010* – a commercial by a graphic design group

*Another example of graphic designers, trained in broadcast television, transferring skills to commercials is described in Chapter 7 by the designer Marc Ortmans. A celebration for 'First Direct' (**a and b**) had computer effects to make a tiny model appear to hold a vast audience. It was introduced by an announcer 'interrupting' transmission (**c and d**). Lasers and spheres (**e to g**) added a futuristic feel*

(c)

(a)

(b)

(d)

(e)

(f)

(g)

11 *The Second Imperative* – an in-house ITV design

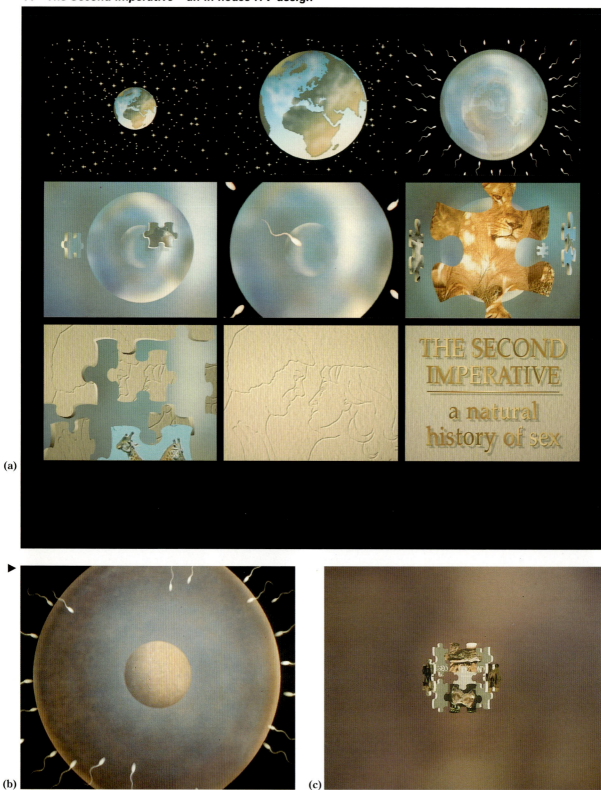

(a)

(b)

(c)

△

(d)

12 Electronic effectiveness

(a)

(e)

(b)

*The highly finished storyboard (**a**) was made on an 'Aurora' paint system. Frames (**b to f**), from the final sequence, show some of the changes described by David James in Chapter 7, in this electronically generated title*

*In Chapter 7 Terry Hilton, of the company SVC Television, reveals electronic equipment used in producing a KP Skips commercial. Footage of figures (**a & b**) were combined with 'Paintbox' backgrounds and others (**c**) were 3D manipulated*

(f) ■

(c)

sional scale development (as unidimensionalized scaling) would eventually lead to more refined mathematical consideration of weighting. However, it is a problem of weighting different factors or components instead of individual items. Here research and theory may complement each other. We could make the scale more reliable or more predictive of some other variable via empirical means. Specifically, the latter can be done by using multiple regression when appropriate scaling models are applied to the data (usually in the linear or linearized situation, or ideally, additive assumption would apply). In the study of the relationship between social support and affective functioning, for example, this is feasible by taking one as an outside variable in scaling the other. After comparing the regression coefficients with other possible weights, Ghiselli (1964) concludes that the beta coefficients are the best weights for the predictor variables as components of the composite measure in relation to an outside criterion variable. As the regression model especially accommodates variables of different dimensions that are orthogonal to (independent of) one another, it represents a convenient, practical approach (in contrast to the theoretical approach discussed in the preceding section) to the issue of unidimensionalization. Here regression serves .the purpose of reconciling the results of factor analysis, which represent different dimensions, rather than individual items that often obscure the issue of dimensionality. As a matter of fact, factor analysis has long been used to derive a set of uncorrelated variables when the use of highly intercorrelated variables may yield misleading results in regression analysis (Kim & Mueller, 1978). This method actually makes the weighting of individual items unreasonable or unnecessary.

Multiple regression has special utility in multidimensional scaling, not just because its logic also underlies factor analysis. It aims at the maximum correlation of the composite variable (to be scaled with a few factors or components) with a specific criterion variable. This is actually the ultimate purpose of scale development, with the understanding of the intimate relationship between scaling and substantive hypothesis testing. The application of a linear regression model also makes explicit the fundamental requirement of linearity for the scales that are simply summed. Moreover, it not only stimulates awareness about item coverage (item sampling adequacy) but also raises the issue of redundancy if items are directly used in the regression equation in place of independent factor scores. It is, therefore, essential to explore the application of such an approach in the interest of both developing assessment of and substantiating the relationships among variables.

The optimal weights obtained by regression methods are characterized by their

13   *Night Network* – an independent group design for ITV

(a)

(b)

(c)

*Mike Bennion of Blink Productions presented the above rough (**a**) when he was commissioned to design a set of 10-second idents for a late-night/early-morning youth series. It is not a storyboard made to represent animation but an indication of the themes he chose for each of the 10 programme sections. His work has a refreshingly anarchic strain and he describes in Chapter 7 his ambition to inject humour and fast pace into this programme styling project. Although 35 mm film was used to shoot all the live-action the final sequences were composed of hundreds of photoprints (**b to d**).*

(d)

3 Three-dimensional computer animation

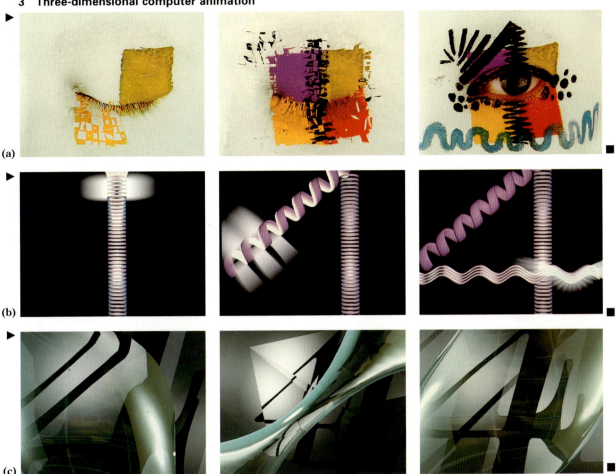

(a)

(b)

(c)

The great achievement of three-dimensional computer animation, in its later stages, is to have become less obvious and in some instances invisible. 3D animation at this level usually involves some original software programing as opposed to reliance on preprogramed interactive devices. The stills on this page are all from a set of 'stings' designed for Channel 4 by one of their own designers, Joe Roman. His roughs for the nine themes on camouflaged numeral '4's were taken to Electric Image

where he worked with some of the operator/art directors on the production and computer realisation of each animation he called 'blips'. Mike Milne was the collaborator on Picasso *(a) and* Drama *(c). Ian Bird worked on* Blip *(b). Martin Foster describes the relationship between computer-programing companies and graphic designers in Chapter 8 and he co-operated in making and animating* Construction *(d)*

(d)

4 Getting a hand from the animator – *The South Bank Show*

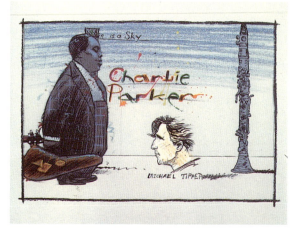

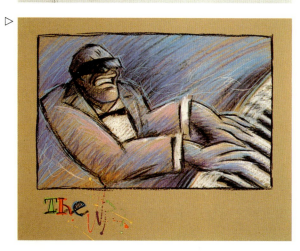

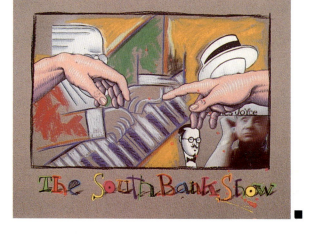

This exuberant piece of hand-drawn animation was Pat Gavin's sixth title design for this programme. He used Hibbert Ralph, the company he had employed on the first series, and Peter Jones Rostrums filmed the titles.

The music was by Andrew Lloyd Webber. Jerry Hibbert describes his work in section 4 of Chapter 8. The five animators were Neville Astley, Kim Burdon, Alvaro Gaivoto, Pat Gavin, and Dennis Sutton

5 A computer program for drawn animation

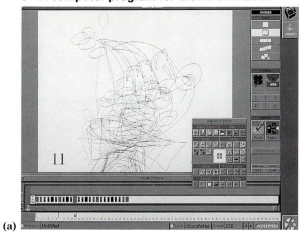

(a)

(b)

*Attempts to make computers speed up the painstaking work of hand animation have been made for a long time. Cambridge Animation Systems approach to this difficult problem is explained in Chapter 8, section 5. A very large program allows endless in-betweens (**a**) to be*

*created with great subtlety once a few drawings are presented and then previewed against any background (**b**), drawn or photographic. Areas can be coloured or 'airbrushed' very fast (**c**)*

(c)

Models and motion

Iain Greenway's title for the BBC drama series A Sense of Guilt *combined a model made at approximately half-scale with motion-control filming. Photographs and captions on pages 114–116 show the shooting process at Cell animation*

Full-size set and live action

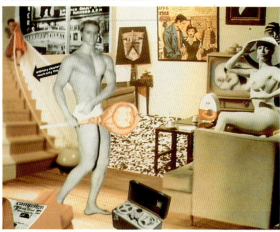

The extraordinary faithful copying of Richard Hamilton's collage is shown in the stills on this and the opposite page. They are taken from Graham McCallum's design for a documentary on the history of television commercials which was entitled Washes Whiter.

Visual stepping-stones

Having to set a mood and convey complex subject matter in only seconds, demanded in television titles, is one of the reasons why graphic designers may decide to use existing icons to help them to communicate. Some instances are considered in Chapter 9. Those here pass the acid test of being applied to an appropriate cause. Graham McCallum of MKD recreated Richard Hamilton's 1955 collage to introduce Washes Whiter and Iain Greenway of the BBC entered Escher's 'Another World' for the drama series A Sense of Guilt

Almost every aspect of the once frozen image was animated in some manner and the continuous action ended in the display of the title among household ideograms. More information on the source is in Chapter 9

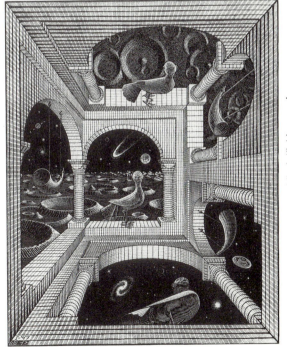

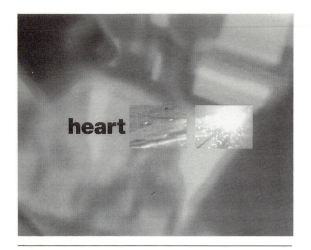

8 Changing patterns in the graphic world

Heart of the matter – *BBC titles by ITV designers*

Here we are at the heart of the evidence of the rapid changes in television production. Two senior graphic designers at Thames Television, Chris Jennings and Jean Penders, explain how they found themselves designing titles for a BBC series.

> After the loss of the franchise in 1991, we had to begin to re-plan our futures, in common with many others of Thames's production department staff. With an immediately noticeable decline in the workload after the announcement by the ITC that Thames would cease to transmit to the London area during weekdays at the end of 1992, we had to look further afield to employ our creative talents.
>
> These changes in broadcasting were not only affecting the Independent Network, but were also starting to rub off on the BBC's working practices, where more and more work was to be put out to independent production companies under pressure from the Conservative government's legislation.
>
> This was an opportunity and a necessity for us. We pitched for the design and production of a BBC1 programme, while still employed as full-time designers by Thames Television, via an independent company, Roger Bolton Productions.

Standards and costs

Working within these new constraints, the foremost being the changing economic realities resulting from the franchise auction, budgets are becoming tighter and tighter. Because an independent production company is working to a fixed budget, and 'graphics' may be at the bottom of the requirements, when there is an overspend (for example on the shooting costs) then money has to be saved elsewhere. Graphic design at that stage tends to become disposable. Quality, anywhere, comes from ideas and attention to detail. If one cannot afford the price, the standards will come down. While in the past a budget actually reflected what the work would cost, a budget now means that when you pitch for a job you have to throw in 'extras' in order to secure the contract from other competitors while the budget remains inflexible and fixed.

Graphic designers and other contributors to programme-making whose livelihoods depend on ideas, are now expected to provide these merely for the kudos. In a freelance world you can cost machine time precisely but what happens to creative time?

Having said all this, *Heart of the Matter* was a fine programme with which to be involved; a prestigious BBC1 current affairs programme dealing with some of the major dilemmas in life such as mental illness, politics, the family and drugs.

In the title we tried to reflect getting to the 'heart of the matter' by using some specially shot footage and incorporating archive material. This live-action footage was juxtaposed with more abstract images relating to time, space and motion. Following the design and storyboard stages the final compilation was made at the post-production facilities house Molinaire, using Quantel 'Harry' to ensure we could layer the various images without losing any of the picture quality.

A footnote: the role of the graphic designer is changing from 'topping and tailing' programmes to taking on the role of the art director, producing and directing sequences. Producers direct and Directors produce, leaning on experienced camera operators and designers for visual and aesthetic guidelines.

Riding the storm

In this changing environment designers must cover a multitude of skills. Those who have been trained and gained experience in the big institutions might be the lucky ones to ride out the increasingly competitive business storm, but what about the newcomers and fresh talent.

It makes little sense to see the high standards of a disciplined medium being dismantled, while other countries, NHK in Japan for example, are trying to emulate the British broadcasting system. Here the government is dismantling and de-regulating, offering us the American example as an alternative.

Credits for The Heart of the Matter
Designers: Chris Jennings and Jean Penders
Client: Roger Bolton Productions
Post-production: Molinaire

The images opposite and below for Heart of the Matter *were made on 'Paintbox' and presented to the editorial team to gain approval for the title. Storyboard and final colour stills are on page 92*

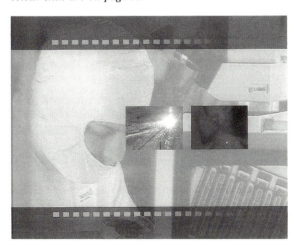

9 Moving the biscuit wrapper

From graphic design for ITV to directing TV commercials

Close-up of the prosthetic head from the final frame

Barry O'Riordan left his art school to join a fairly young Thames Television at the time when some of the most memorable programmes were in production. Jeremy Issacs was directing the ambitious series *World at War* and Barry was assigned as graphic designer to another 12-part documentary series, *Hollywood*.

His experience in every possible production technique, and a collection of design awards from BAFTA, D&AD and the RTS led him to his present post as Art Director at the Moving Picture Company. MPC was founded in London in 1970 as a commercials production house and it expanded rapidly into post-production work, opening the first video tape editing facility for the growing television advertising industry.

Among the most innovative companies of its kind, The Moving Picture Company built up a range of equipment to serve graphic designers for broadcast and commercial work and this presently covers five editing suites, (three of these are component D1 and composite D2 domains), Telecine transfer suites including 'Matchbox' for perfect matching shots, and Ultimatte 5, 3D and 2D computer graphics departments. Here the hardware includes 'alias' 3D computer-imaging systems with the super-fast Silicon Graphics 4-D-120 renderer, three Quantel 'Paintbox'/'Harry' suites with 'Encore'. There are also a video rostrum camera studio and a computer-controlled motion-rig facility. Their reputation for special effects has become international.

Each facility house has its own recipe for the mix of software and hardware; the problem for graphic designers is trying to keep abreast of who has what, as well as knowing how each device in the chain can be used.

Clearing the production path

To relieve this headache The Motion Picture Company was among the earliest to employ graphic designers and art directors as full-time staff members. Co-operating with advertising or TV broadcast clients, the in-house graphic designers' experience of translating images to animation can guide whole projects, or as in this case, work with many others to control the complex production path and realise other designers' concepts.

The making of the 'MC Classic' commercial for Fox's Biscuits shows the composition of skills and the role of the graphic designer in a facility house. MPC report:

The creative idea behind this 30-second commercial is based on the popularity of the 'rap' style where the traditional lyric is replaced by a rhyming chant or message. The visual style of videos of such music is characterised by energetic dance routines cutting between locations and costumes and lighting.

Our brief from the advertising agency told us 'MC Classic' the central character, was to become the biscuit. His body would appear as the reflective printed wrapper, whilst his head, neck and the top of his chest would be the chocolate covered biscuit. 'How could this be achieved in a visually convincing way?'

Three-dimensional 'Alias' computer-generated animation seemed the perfect solution to create an ultra realistic model of the pack with reflections, shadows and highlights. However, it would not be suitable to animate the range of expressions in the face. For this reason we opted to make the live-action resemble a computer graphic rendering by using a thin rubber prosthetic capsule to cover the head, neck and chest.

The costume designer was then briefed to produce outfits in the pop video genre, allowing for plenty of movement and echoing elements of the product and the colours of the wrapping. As for 'MC Classic' a very special outfit was conceived for him. As his real-life movements would form the basis of the computer animation, it was necessary to construct a shape which was similar in its proportions to the actual rectangle of the product's wrapper, but still allow his movements.

A concertina-pleated floor-length piece of fabric was produced, dyed a similar shade of blue to the actual wrapper. This meant that any interactive lighting effects would match in post-production. On top of this was fitted a rigidly starched off-white collar which looked like the torn-open inner wrapper in real life. It was necessary for the edges of this collar to be clearly defined, forming the edge guide in post-production for the computer animator to follow; everything below the collar would be post-production animation and above would be live action. The arms were left uncovered to be replaced by computer-animated wisps of the wrapper, designed to follow the live-action arms.

Having rehearsed the carefully choreographed routine, fitted the costumes and designed the wigs, the live action was shot on 16 mm film to reinforce the 'pop video' style. A 30-second edit was then approved, permitting the transfer of live action to D1 tape and feeding frame-by-frame into the 'Alias' system.

To aid the three-dimensional modelling and animation process, a separate video camera on the live-action shoot had recorded MC Classic's position from the side wherever our main camera had shot him from the front. We then cut this footage to match the approved film cut in order to have accurate reference of his whole body movement.

As each shot was completed on the 'Alias' computer it was crudely mixed over the live-action to check position and length before being laid off on to D1 tape as matte and master runs, as well as shadow and highlight passes.

Everything was finally assembled in 'Paintbox' with 'Harry' and 'Encore'. This enabled subtle effects such as softening and motion-blur to be introduced, blending computer-generated graphics with live action more convincingly.

Credits for 'MC Classic' commercial
Client: Fox's Classic Biscuits
Advertising agency: The Graham Poulter Partnership, Leeds
Creative Director: Pete Camponi
Art Director: Mike Clough
Copywriter: Ventnor Brewer
Agency Producer: Bill O'Brien
Production Company: The Moving Picture Company
Director: Barry O'Riordan
Producer: Hilary Davis
Post-production facilities: The Moving Picture Co.
Computer animation: Gareth Edwards
'Paintbox'/'Harry': Paul Round and Harry Jarman

10 Into the future – to *2010*

A television commercial by Ortmans Young

A superbly researched and very detailed account of this 90-second commercial was prepared by Peregrine Gibbs, a graduate of Bournemouth Polytechnic studying Communication and Media.

His dissertation – '2010 – from conception to reception' would fill this entire book. He discusses every possible aspect of the project, from the formation of the advertising agency (who commissioned Ortmans Young, the graphic designers/producers) to a study of the results of the commercial's effect on the audience.

This review demonstrates the need for, and the logistics of, an advertising agency working with graphic design and production expertise to translate their message to effective moving images on-screen.

First Direct/2010

The launch commercial for a 24-hour telephone banking service, to be called 'First Direct', required Ortmans Young to draw a great deal upon their experience of designing for broadcast television. The theme of the commercial was to be set in a *War of the Worlds*/Orson Welles mould – aimed to convince the audience that they were experiencing a transmission from 30 years ahead. They were to create a mysterious event which might provoke press and public comment.

The designers enjoyed a close working relationship with the advertising agency Howell Henry Chaldecott Lury, (HHCL) and their client Midland Bank. This trust had the advantage not only of ensuring imaginative development of the concept but permitting rapid decisions to be made at each stage.

From storyboard to set

During the early stage of translating the agency's message the concept of the commercial was changed from the idea of a futuristic experimental video to a more comprehensible notion of an 'anniversary party' in the year 2010. Design and pre-production phases had to move quickly to ensure completion by the transmission date.

A 10-storey scale model of an atrium was built in sections to a height of two metres, complete with a complement of a thousand model 'spectators' glued into the balconies. 'In the studio it looked like three shower cubicles jammed with people', said Marc Ortmans. Peregrine Gibbs commented 'The model had to create

Ortmans Young clients have included Philips, Siemans and Courrèges. The rise and fall of the city blocks in this commercial showed how Neefax grows with and supports large and small firms

specific data base in place of theoretical application. They are also featured by specific objectives and corresponding techniques (Horst, 1966). Yet, the results obtained from individual data sets are subject to explanation, evaluation, and modification from the perspective of established theories. Without appropriate empirical basis *and* adequate theoretical preparation, it is a real danger to use any computational procedure for scaling purpose. By presuming every indicator to be equal, which is the usual case in summated scaling, for example, it is very likely that we will lose important information on items (factors) that have relatively small variations but that in real terms mean a lot to human conditions. Even standardization of the individual items will not guarantee the suitability and utility of a scale, since the role of each item is determined by complicated considerations, including, for example, the degree of sampling adequacy. Most importantly, psychosocial domains tend to be multidimensional rather than unidimensional, thus the application of unidimensional scaling techniques could be very problematic indeed.

The use of the criterion variables is an important strategy in statistical analysis, and in scale development in particular (e.g., criterion-related validity). By relating to the criterion measures based on our theoretical beliefs in their associations with the constructs to be scaled, we attain an important passageway to unidimensionalization. In general, multiple regression will yield optimal weights by using such criterion variables. And the statistical procedure has with it serviceable means for searching for violations of assumptions such as linearity and normality. This is important for scaling based on individual items. Once serious violations are detected, remedial measures can be taken to transform the data to meet the requirements for such analysis. Nonlinear regression analysis may also be considered. These techniques, therefore, seem to outclass the conventional item scaling procedures. Scale development based on principal components or factors, however, will sidestep all these issues.

The above strategy has a practical orientation since it targets specific criterion variables, which are usually of particular research or practice interest. To try out such an approach on the empirical basis, careful consideration of the criterion variables needs be taken for each of the key constructs to be scaled. Usually linear multiple regression is the simplest for scaling purposes. The analysis of residuals can be performed to assess the goodness of fit of the regression model. To avoid the interaction effects among the components, factor solutions from orthogonal rotation can be used for regression analysis. This is indeed the advantage of the component/factor models over regression analysis using individual

an illusion of enormity – and this demanded an intricate contribution at the post-production stage.'

The model background scenes were then shot stop-frame using the motion picture control rig at Moving Picture Company.

Lasers, light spheres and other displays were constructed using computer-generated animation and this was entrusted to Craig Zerouni and Computer FX. His approach to his clients is 'What is your dream? Let us help you make it into a reality'. In this case Computer FX developed software to transfer the digital pictures on to the higher resolution of 35 mm film, using technology to enhance the production of the design. Zerouni felt '"2010" worked better than I expected . . . and served to be as much of a challenge as it was a chore. At times it was great fun'.

The motion control data was transferred from the camera computer in order to register the computer animation within the atrium for later compilation. Real dancers and spectators were filmed on the same motion control system to add scale and reality. The budget limitation required two dancers to play the roles of nine, and the spectators' scenes were achieved with 10 extras and numerous costume changes.

Once shooting was completed the 35 mm film was transferred to digital video and painstakingly edited using a 'Harry' digital editing system over an exhausting nine-day marathon.

Finally the soundtrack was composed to the finished video edit and completed with just a couple of hours to spare. The whole project had progressed from the first meeting to transmission within 12 weeks. As all similar tasks, it represents a large team effort, a great deal of professional commitment and planning.

Marc Ortmans concluded:

The greatest reward is discovering how the audience reacts. Judging by the thousands of telephone reports, taken by ITV and Channel 4, of aliens invading living rooms, it succeeded in its objective.

Credits for 2010
Agency: Howell Henry Chaldecott Lury
Design directors: Marc Ortmans and Haydon Young
Model-maker: Mike Kelt of Artem
Motion-control filming: Peter Tyle/Peerless Camera Company
Computer animation: Craig Zerouni/Computer FX
Music: McCormack

11 Production decentralised: distribution worldwide

The second imperative – a documentary series

HTV (Harlech Television) retained their franchise for Wales and the west of England at the end of 1991 and, like other ITV contractors, they are now enjoined to make more programmes with outside producers.

HTV has recently been commissioned by Genesis Film Productions to make the title sequence for *The Second Imperative*, sub-titled *A natural history of sex*. The in-house graphic design unit, headed by David James, was sub-contracted by Genesis to carry out the graphic design work for the whole series. This resulted in the programme originators, the designer and the production services being widely dispersed. Working in this way is bound to become more commonplace and this experience is described here by David James.

(Since this account was written the series title has been changed to *The Sexual Imperative* and the design and the animation has been changed by the designer as follows: instead of the incised lettering the pieces show the female and male gender symbols and these mix at the end to the new title lettering.)

This title is fairly representative of its genre. The brief was a familiar one, an opening sequence which would serve two functions: first, to intrigue the viewer with its narrative use of imagery and second to impart a visual précis of the subject matter to follow.

Some commissioning producers undervalue the significance of the opening sequence. Titles and credits are more than the equivalent of the covers of a book; the opening sequence should be a foreword. Like a concert, a television programme is a one-off event, competing for the viewer's attention at a particular moment in time. As an overture to a programme, the title should establish the subject, style and mood to follow. Its relevance should be based on forethought, integrated into the production and formulated at the planning stage, not as an afterthought made with the remnants of the budget. Most designers remember an idea on the drawing board which was stillborn on its realisation. In most cases it was the result of an incoherent and ill-considered brief; the director all but says, 'I'll know what I want when I see it.' In this instance I was fortunate. I found I was working with a director,

Dr Clive Bromhall, of Partridge Films, who was not only a specialist in the field to be covered but who could also articulate a sympathetic and critical awareness of the visual problems involved.

The project started with a verbal briefing for the draft storyboard. It was at this juncture that I faced the primary obstacle in the preparation of the sequence: that of the distance between its creative contributors.

As the television industry becomes more decentralised with specialists based in separate offices and facilities, clear rapid communication is of major importance. In the case of this series, the titles were to be commissioned in London, designed in Bristol, and realised in Reading and London. The fast exchange of visual information is now accomplished from our HTV Bristol studios by the use of A4 facsimile sheets and when accurate original artwork is required, discs of the type used in desktop systems like the Apple Macintosh. In this instance, however, I decided to prepare more finished frames, which would convey fairly closely to both director and animator how the final result should appear. The images were produced on the 'Aurora', an American painting and animation system, and photographed with a video stills camera and then mounted as prints in position for the storyboard.

Brief to storyboard

The draft storyboard is not a precious stage of the process but a very significant one – the first assembly of ideas, a structure which can be built upon, modified or rejected. Clive Bromhall had several constructive criticisms of my interpretation and we began a process of elimination and refinement over the course of two further storyboards until we considered that the narrative had been pared down to a simple sequence of events, linked in a continuous flow.

The next stage involved the 3D animation. Owing to full bookings on the 'Aurora' at Bristol design, this was undertaken by an HTV company, Headline, at Reading. They were supplied with the storyboard in animatic form on Beta and as a hard copy in the form of photographs.

The animation divides into three sections: the tracking-in on the earth, the transitional mix to the sperm entering the ovum and, finally, the orbit of the jigsaw puzzle pieces.

The earth shot was relatively simple. A NASA photograph was texture-mapped partly around the front of the face of a sphere.

Two-dimensional animation formed the transitional section and this was achieved by Graham Young through the use of Quantel 'Paintbox' cut-outs animated and rendered in multi-layered passes on the 'Harriet'. We were concerned about the problem of achieving the right effect for the gentle abstract background but Graham solved this by diffusing light into a camera lens through a glass fruit bowl slowly revolving on a record turntable. Ten seconds of footage was flipped, tumbled and played over itself into 'Harriet'. Simple but effective.

In the final section, one jigsaw shape served as a master for all the pieces, to be used horizontally or vertically. Animator Helen Hywel plotted this shape on the 'DGS' animation system, using 200 points, to give a smooth outline when viewed close to the camera. A 3D model was then extruded and all the pieces checked in their final interlocked position. The title lettering was then incised on the back of each puzzle piece.

From this, a new model of each piece was built. Again, the pieces were assembled as a total 3D model to check that the lettering – cut to a depth of 75 per cent of that of the puzzle pieces – would be legible in its final form and with the set lighting positions.

The final phase involved the placing of moving images on the front face of the puzzle pieces. Richard Gardner achieved this on the Quantel 'Harry': each picture was represented by a three-second moving shot which was flown in an orbit to match exactly that of the piece on to which it would be placed. Mattes supplied by the animator allowed each image to be keyed precisely in to the shape of the puzzle piece. The final result was recorded on to BetaSP.

It is standard practice for the music to be the starting point of a title sequence. In this instance, the visuals preceded the music. A copy of the finished sequence was sent to the composer Terry Oldfield. Terry wrote the music to match precisely the action and mood of the sequence. The result was startling: a new life was breathed into our images and the original character of the titles was changed.

Credits for The Sexual Imperative
Production company: Genesis Film Productions
Graphic designer: David James,
Head of Graphic Design HTV
Animation: Headline Video Facilities, Reading
Music: Terry Oldfield

12 Electronic effectiveness

Graphic effects in a television commercial

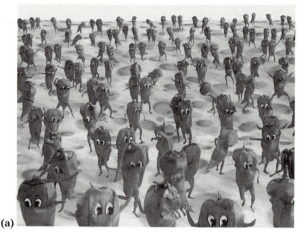
(a)

Terry Hylton acquired the ability to transform his own visual ideas to the screen through his extraordinary enthusiasm and knowledge of electronic equipment as it evolved around him during his years as a graphic designer at the BBC.

Now a Creative Director at SVC Television, he has the reputation among a large number of advertising agents and his fellow graphic designers of being a 'whizz' at interpreting a storyboard and knowing just how each piece of equipment can transform the design to the screen – like a conductor with a splendid orchestra.

In Chapter 8, section 8, he describes post-production's current technology and the difficulties that confront designers when realising their designs – at a price and on time. Here he shows the part SVC played in a KP Skips commercial.

In these days of tighter budgets and ever shorter deadlines it was a refreshing change to be involved in a commercial that was a bit different.

The first thing that made this KP Skips (a snack food) television commercial unusual for me was that I found myself *'pitching'* to the advertising agency and the production company with my proposals for the effects in their storyboard. The product *'melts in your mouth'* so a visual effect was needed to put this across and create something memorable.

(b)

Electronics enable SVC to create an impression of a cast of hundreds when only a very few characters were used, (a), and at other points in the sequence (b, c & d) apparent swift movements of the camera (whip pans). A 3D computer made the distortion for the end shots

Micro-tricks

A test I presented did the trick. After two subsequent meetings I found myself at the first day of shooting. The concept of the commercial to launch the new product was that three flavours would be created by thousands of microscopic characters dancing and fooling around on the surface of their respective Skips. The budget did not stretch to thousands of extras. Material had to be filmed in such a way as to allow the dozen or so chillies, spice pots and spare ribs, which we did have, to be replicated.

Another aspect of this project that was out of the ordinary was the fact that the traditional role of employing a film editor to produce a cut version was bypassed. The director, Max Vadukal, worked with Tom McKerrow, Senior Editor at SVC and cut it on videotape.

(c)

(d)

(e)

Art Direction: Terry Hylton
Director: Max Vadukal
Videotape Editor: Tom McKerrow

Once this cutting copy had been approved, the job of replicating our flavour characters could begin.

One shot was created in a digital suite using footage of the artistes filmed against white and then reduced and positioned in a DVE machine against a 'paintbox' (1) background. Shadows were also added to mimic the true shadows that had been present in other shots. While this operation was taking place – it actually took three to four hours to finish – 'Harry' was already working on another scene that involved frame-by-frame painting to build a close-up of more chillies. It is not that rare nowadays for both areas (DVE and 'Harry') to be used in tandem but it was to continue for three days, allowing a fair amount of experimentation and enabling us to get the best out of the respective pieces of equipment.

'Painting' the pack

The main body of the ad featured three sequences of the different characters causing mayhem. These were linked by whip pans that had been shot especially to create the impression of moving away from the surface of one product to another. (2, 3 and 4).

No commercial would be complete without a packshot and, just to make sure that everyone got the message, this ad had two! In order to arrive at the first in an entertaining way the storyboard called for an infinite zoom out from one spare rib to reveal the whole Skip sitting in front of three packs. This was created as a 'nested' zoom reducing the first image by 50 per cent in one channel of DVE, then extending it outwards to fill the screen with 'Paintbox' through a second channel. The two channels were reduced together and more painting extended outwards until the final frame came into view. The resulting sequences were speeded up in 'Harry' to fit the slot in the cutting copy.

3D manipulation

A hand then picks up this solitary tasty morsel. A 'Peewee Herman' type character proceeds to swallow it. He appears to have rubber arms and legs and visually melts as the Skip is digested. The footage of him swallowing was manipulated in 3D. His limbs were further stretched and squeezed by computer to create a very amusing penultimate shot (5). This mixed through to the real packshot which was animated to mimic the melting that had already been seen at the beginning of the commercial. With the audio recorded and dubbed and a few final tweaks to the overall colour, the whole commercial had been post-produced in just four days. the client was delighted, the campaign was a resounding success and as I finish this report I have just heard that a follow-up has been commissioned.

13 Mikes big London night out

Designing for the teenage sector and LWT

Mike Bennion, a director of Blink Productions Limited brings a youthful, bravura style to the television screen echoed in the print world in *Face* magazine and the work of other new generation groups – Why Not Associates, Malcolm Garrett and others.

Leading, or translating, the contemporary mood of the vast potential youth audience for television has been a task for each generation of television graphic designers. Widespread accolades of Bennion's approach in the professional press already show his influence in the early 1990s. He describes a commission from LWT as follows:

> Night Network was a brave attempt by London Weekend Television to corner the youth audience on Friday night/Saturday morning. The show had been running for one season already but the three-hour programme had not really gelled as one programme.

Ten-second sequences
The series producer felt it needed some visual glue, and that's where I came in. They had broken the programme down into 10 bits, called, for example, *Small Screen* (about TV reviews), *Press Gang* (about newspaper gossip), MBTV (Mick Brown, a DJ spouting-off), and so on and so forth. My bits took the form of 10-second title sequences for each of them. The programme itself was fun, light and frothy so I aimed to reflect that. The music already existed for the stings they had used previously and I was obliged to use it, which often dictated what the visuals looked like. I chose to have the kids in each sting and have them do something loosely connected with the titles; so for example in *Small Screen* one kid puts a television on his head, another jumps up and down on sets in the background; for *Pillow Talk* two have a pillow fight. As simple as that. For MBTV I had a kid yawn for 10 seconds. I'm still amazed I got away with that!

I kept all the typography very small purposely as everyone else in TV graphics was making type full-screen at that time. The colour scheme was black and red to pick it out from the main body of the programme and give a uniform identity.

Having fun
Technically, it was a one-day shoot in a tiny studio with about eight kids, shot on 35 mm film. The selected rushes were transferred into hundreds of black and white prints which were then scribbled on, overlaid with acetate photocopies and generally abused. They were all re-shot under a rostrum camera using a lot of animation cycles and repeated moves. Captions and other simple animations were added by being fed into 'Harry' and 'Paintbox'.

I chose this as an example of my work because it was the most fun to do and I think it showed in the final project. A postscript to the whole job was that a new programme director was brought in at the last minute. He decided my bits were too much of a coherent identity for a youth programme and proceeded to pump bright day-glo colours into them, stretch, distort and re-cut them. The end result was still a coherent identity but now it looked 'shit'. Fortunately, hardly anyone stayed up to watch *Night Network* anyway and the series was pulled after a year or so. That's showbiz!

Credits for Night Network
Art Director: Mike Bennion
Producer: James Studholme
Camera: Les Nutt
'Harry' operator: Mike Magee/Framestore
Series Producer: Jill Sinclair ☐

Chapter 8   Services for the Graphic Designer

1 The modelmakers and visual effects

Model animation has never 'gone out of fashion' with graphic designers. Their association with model-makers and special effects experts has been long and sustained.

The work grows from a shadowy area where two- and three-dimensional presentation overlaps; where a knowledge of construction in precisely the right material for filming and where the use of armatures allows controlled movement of models to be photographed frame-by-frame.

The BBC has a London-based Visual Effects Department which at its height employed about 80 people. From there they can supply anything from the most intricate miniature to full-scale firework displays. Graphic designers at the BBC could have asked them to create mosaic floors for a Roman bath or the pyrotechnics used to make the title for *Prince Regent* (1979). They have a large stage, where models can be lit and shot in film or video, and facilities for life-size sculpture and moulding at the Visual Effects Complex at Western Avenue. They also work for the scenic designers, making props and creating effects from floods to infernos.

There are many independent model-making companies who have developed a special knowledge and understanding of the particular needs of graphic, film and video work. They are commissioned by designers in all sections of the industry.

The latex loaf

Sue Lipscome runs the strangely named model-making group, 'Codsteaks', in Bristol. She and her team delight in preparing routine and bizarre models for graphic designers and animators from the BBC, ITV and many of the independents.

Graphic designers are, according to Sue's experience, fastidious and demanding clients. The models they commission are seen in close-up by millions of people and the modern television cameras have a penetrating lens. Precise finish and materials that imitate the real world in perfect colours are demanded.

One of their customers, and by chance close neighbour, is Aardman Animations who set them the task of making the world's first 'trampoline loaf' for a stop-frame animation television commercial for Lurpak Butter. Moulding the bread in latex was not exceptional but fitting a complex 'rig' inside, to control the surface during filming, is typical of TV model requirements. Sue's training was in 3D work at Bristol Polytechnic and later a year's postgraduate study in Theatre Design at the Bristol Old Vic. She then started Codsteaks. (Her explanation of the name is that she was merely shopping for them on the day she and her original partner decided to set up the business. She later thought it might invoke 'COD' – cash-on-delivery – in the unlikely event of any customer being slow to pay.)

They did work for graphic designers but after four years Sue felt it was vital to discover the business from the inside. With great enterprise she joined the BBC Visual Effects on a short-term contract and became for 18 months a 'spy'.

The experience gave her both greater insight and a lot of information on the use of materials and, as she says, 'greater credibility as a woman in an unusual career.' Six years later that singular attitude has created a list of credits for model work on television programmes like *Little and Large* for the BBC, *The Garden Club* for HTV, and many others, as well as commercials working with graphic designers for Pirelli tyres and Cadbury's Crunchie.

Filming food under studio lights, often over days or even weeks, means that the most commonplace items, from rolls of bread to bowls of very drinkable soup, have to be conjured from the most unlikely materials.

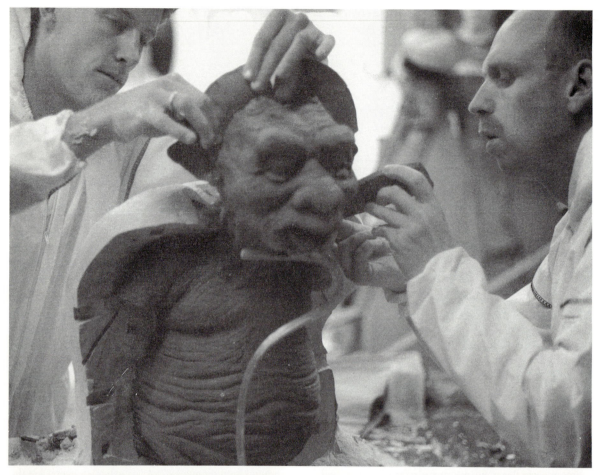

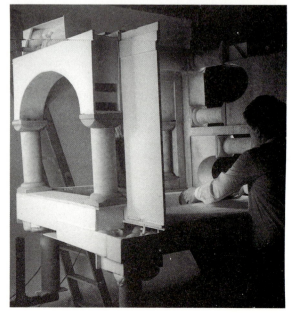

Making dreams come true

Alan Kemp is another model-maker whose work has been almost exclusively for television graphic designers during the past 12 years. He says 'They seem to be people who spend most of their time dreaming up strange things for me to make !' His obvious enjoyment in translating their daydreams into wood, metal, plastic, stone – and almost every other known material – has given him an unusual career.

Kemp studied sculpture in the School of Art and Design at Wolverhampton Polytechnic. For 10 years he had been an art teacher when an acquaintance from college days, who was then working as a graphic designer at the BBC, asked him to prepare 3D magnetic symbols for the weather map. (Archaic now but used every day in some some stations' transmissions throughout the 1980s!) This modest job was constructed, silk-screened and delivered within the brief time available. Next he was making stone lettering for the word 'Holocaust' to improve on the too obviously polystyrene version that had met with the hostility of a presentation director.

Within less than a year his workload and a list of satisfied clients led him to become a full-time graphic design model-maker, 'by appointment' to the BBC and Thames Television and other clients. His work on the large model for Liz Friedman's titles for *Chronicle* has been described in Chapter 7.

2 Motion-controlled rigs

Using a movie camera to film a model was routinely exploited in the early days of Hollywood. Anything from a scale model representing a huge ampitheatre complete with animated crowds (*Ben Hur*, 1925) to a burning skyscraper was constructed and filmed. The shots were then back-projected, or used via the aerial image of a rostrum camera. Such models saved large amounts of money on scenery, or to carried out effects which could not be achieved by any other means.

Graphic designers in television have borrowed, and in many cases extended, this area of production. In the early years a film camera, sometimes using single frame, would be mounted on to a smoothly moving dolly, often guided by floor tracks to shoot either a model or objects in a studio. Re-shooting, or trying to make more than one pass for making mattes, was almost impossible because of the imprecise positioning of both camera and object.

By the early 1980s camera operators and designers in many parts of the world were beginning to harness computer-control to cameras and mount these on rigs whose fractionally small movements were also controlled and recorded by computer. This has enabled film or video cameras to shoot any sequences, no matter how complex, and repeat them at will even days later.

Working with models forms a large percentage of television graphic design. The BBC has for many years had a model-making and special effects department to support the set and graphic designers. Building models to be used in studios that can, lit well, be often capable of precise movement, is essential. The photographs on the opposite page (above and far left) show BBC modellers in the workshops and in a film studio

The photographs, left and right, are of a large model made by Alan Kemp for A Sense of Guilt – *a BBC drama series designed by Iain Greenway – and they show preparation where a figure representing the main character is adjusted before the motion-rig filming at the London company Cell Animation*

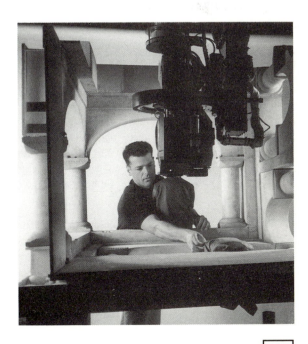

Model work demands very high-quality lighting equipment and skill from the camera operators. Colour stills from this shoot appear on page 102

The Moving Picture Company built one of the very first of these units in London and they have recently upgraded it.

The motion-rig operated by Doug Foster at Cell Animation in Charlotte Street was used in the making of the titles for the BBC programme *Chronicle* and this collaboration is described in Chapter 7.

This was designed and installed about five years ago by Mark Roberts as a ceiling-mounted computer-controlled unit with an S35 Mitchell film camera. Foster explained that the concept of shooting models in this way has been greatly enhanced by Cell's in-house Quantel 'Harry':

> There is real excitement in shooting jobs on the computer rig and then composing and editing them on 'Harry'. They are made for each other. The ease of generating mattes and masters within the motion-control rig with separate film passes is superb. Rather than suffer the ulcers and mental disorders so common when trying to compose things on one piece

of film in the camera – including type – we now shoot everything we possibly can as separate elements and then we put them together on 'Harry'. This way of working has evolved and is now the accepted system – far more certain and relaxed, as well as giving scope for alternative ideas to be tried.

Our major clients include the Presentation Group and designers from the main Graphic Department at the BBC, Lambie-Nairn & Company, English and Pockett and McCallum Kennedy D'Auria among the independent design groups, and our very close neighbours at Channel 4 are often here.

As I trained and worked as a graphic designer I feel I can understand their approach very clearly. The earlier they come to talk to me the more I am able to guide them with technical aspects of shooting which I have managed to gather over the years and the more chance we have of producing what they want. Directors with filming experience, in commercials or feature films, tend to work things out for themselves whereas designers bring their storyboards. We can then work out the best way to tackle things and find out anything they may have doubts about. This way we have built up a lot of trust with designers and they seem to come back!

Liz Friedman came to me to talk about designing *Chronicle* even before she presented her storyboard and when she had three separate ideas in mind. She wanted to find out what would be easy, rather than a nightmare to shoot, and that I was confident with her concept. Later we discussed the size of the model we would need and how we would shoot the sequence. It is at this stage that I have a chance to say that although an idea might look good it would cost a fortune to film or that it would be ideal for motion-control. We can save an enormous amount of trouble.

When everything was ready and the model had been delivered we used a periscope with a very wide-angle 12 mm lens to capture the whole word from above with the limited ceiling height and filmed at one frame per second with the camera moving all the time so the motion-blur looked absolutely natural – as if it were shot at 24 frames a second. The lighting was a simple five-kilowatt lamp with the Fresnel lens removed to give sharp edged shadows – like sunlight, and there was a dusting of smoke in the studio to give a feeling of aerial perspective. The aperture of the lens was f22, very tiny, to give great depth of field so everything was in focus as it would be if you were shooting from a helicopter. We had to shoot twice the length required, to prevent the slight strobing which occurred in the tests as the movement was so fast. We made every frame into a field and then reduced it to make it flow more smoothly – that is 50 frames a second in PAL. The lens was left unshaded to allow lens-flare, which you get from the sun, again to appear more natural.

In commercials we have been using motion-control a great deal – on some occasions shooting the models with the rig and putting these into computer-generated backgrounds and on other occasions computer-generated characters have been matched to model backgrounds.

Motion control is certainly a major contribution to the many ways animation is used in television and it will continue to be used alongside them, whatever other changes occur.

A computer-controlled camera suspended from the ceiling is filming the model made for Chronicle with a periscope lens at Cell Animation

3 3D computer animation

An article in a magazine, *Design Issues*, published by MIT School of Design in the University of Illinois, Chicago in autumn 1991 was entitled – *"Real" Wars: Aesthetics and Professionalism in Computer Animation*.

Written by Lev Manovich this made an intriguing claim. The purpose of advances in computer animation were not primarily made to create better illusions, make images ever more lifelike, more persuasive to the viewer – rather:

> they allow the designers to signal their professional status, thus serving as tools of competition within the industry. The struggle to simulate the 'real' masks another struggle – the real war for professional survival.

Manovich went further:

> Today, the importance of technical standards in keeping amateurs from entering into professional markets can be seen even more clearly in the video industry ... The separation between amateurs and professionals in computer animation is first of all achieved through the aesthetic standard of smoothness.

Whether or not these were ever the aims of Electric Image, a Soho-based company founded by Canadian-born Paul Doherty, they have excelled at 'painting by numbers' for the past 10 years. They are archetypical of those who have purchased equipment and software, and have been at the forefront of providing computer power to graphic designers in broadcasting and television commercials, helping them to translate their story-boards into three-dimensional animations.

A recent commission, from the Danish telecommunications company Kommunicator, took over nine months to complete. They worked with the graphic designers in Denmark on a programme for television on the seven wonders of the world. The skills they apply have the aura of an eighth.

The constant updating of knowledge and equipment in these companies has developed at a staggering pace. The phase of brashness and detectable computer images has largely passed – effects and the construction of objects and the rendering of surfaces are now extremely subtle. It is often impossible for even the

items. For the purpose of scaling and model testing, the standardized predicted values of the criterion variable should be saved as the scores of a new unidimensionalized scale formed by the extracted components or factors.

It should be noted that different dimensional analysis procedures (e.g., factor analysis and cluster analysis) will render different results. Therefore, a universal procedure of unidimensionalization is more important and useful than the confirmation of a particular latent pattern. However, we have not scrutinized the difference between principal component analysis and factor analysis for the purpose of scaling. Although either model will do for data reduction, generally principal component analysis rather than the more accurate common-factor model is favored for extraction and dimensionality clarification. There are three reasons for this. One, the theme here is unidimensionalization; that is, the purpose is to explore some viable approaches to the general measurement of theoretical constructs instead of elaborating on the latent factor structures of the specific data sets. Two, without requiring imposing "what some may consider a questionable causal model" (Kim & Mueller, 1978, p.20), principal components analysis is the only means by which exact factor scores can be obtained (Norusis, 1988). And three, some factor models entail multinormal distribution of the variables, which has been frequently violated in practice, whereas principal components analysis does not hold such stringent requirements (Morrison, 1990). As Kim and Mueller (1978, p.72) write, "if the objective is some simple summary of information contained in the raw data without recourse to factor analytic assumptions, the use of component scores has a definite advantage over factor scaling." Particularly, when the real data do not possess a plain common factor structure or coincide to some a priori conceptual scheme, this approach is most practical to summarize the empirical information by unidimensionalized scaling. This is in fact the case in most studies.

In terms of the state of the art of scale development, the CAUS has some fundamental implications with standardized scaling methods formulated to establish comparability among various measures. In contrast to the variety of elaborate schemes of theorists in content analysis and the advanced techniques of methodologists for confirmatory factor analysis, the new approach represents a significant step toward the standardization of research results. It is unfortunate that even psychometricians would allow tremendous variation in practical scale constructions. But now, at least some sort of standardization is possible. Chen's work shapes some principles in scaling process starting from the very first step. These include the clarification of specific research objectives, basic logical

Martin Foster's 3D animation knowledge brought to life Joe Roman's design for a promotion sting for Channel 4, called Topical

most experienced eye to detect 'computer graphics'. As an element of 'tools of competition' they have defeated their objective by becoming invisible.

Martin Foster is one of those graphic designers who has contributed to this field and he describes his work at Electric Image as follows:

> Most of the operator/art directors at Electric Image are graphic designers. Other companies will split the work into design direction and technical production but we each undertake all aspects of the process. We listen to the brief from the outset, determine the best route, give the estimate and, crucial to this type of work, plan the schedule to ensure the availability and capacity of the equipment is there to meet the deadline.
>
> We will then, as technical operators with 'hands-on', build the wireframe models, create the animation and render the colour, texture and lighting to the final tapes – every aspect.
>
> If new software is required, for example a different model is needed for every frame, we can call on the five programmers who use C-language to do this. Very simple programs I can write myself.

Working with the graphic designer

A graphic designer will come in with a storyboard and the first task is to discuss exactly what the end result is to be, then assess the equipment requirements. Invariably the designer is asked to make drawings of the type, objects, figures, etc, from which the computer models will be built on-screen. These will be black-and-white keylines of plans and elevations. (Occasionally we can use a 3D model or the actual object.) We rarely 'grab' the drawings with a video camera as we can be more precise, emphasising or excluding details which the camera can not. Some computer houses employ 'modellers' – it's a highly specialised part of the system. This is likely to be done on an Iris work station at present although we are soon to change to other Silicon Graphics stations called Indigos. Iris has been a very powerful computer for many of our purposes – modelling, animation and rendering – for about 10 years. The Iris interactive wireframes can be moved with dials or with a mouse.

We maintain that the graphic designers who are our clients need only be aware of these techniques – we are here to do this. They have come to understand what we can do and to trust us.

More and more sequences are 'mixed media' – it's a long while since we did a wholly computer-based graphic piece.

You can compare three-dimensional computer-graphics with airbrushing. When I was at art college everybody wanted to be the greatest airbrush artist. I was taught by Harry Willock. He did the artwork for *The Butterfly Ball* – a book designed by Alan Aldridge. Even then we realised that airbrush could combine with freer drawing and painting. No need to be a purist – just get out ink, brush or pencil and finish. It's the image that counts – not how you've done it. It was a terrific lesson to accept that you could mix techniques.

This applies to computer animation. Why spend ages digitising type when you may in-put lettering in an edit suite with a character generator? Why spend time doing things which could be easily accomplished on 'Paintbox', 'Harry' or even live action? Why create absolute 'reality' if it's only going to look real? The only reason is if you are going to achieve something which cannot work any other way. People are beginning to accept 3D computer graphics as simply another tool.

3D animation for Channel 4

Joe Roman is a graphic designer who has worked on presentation material at Channel 4 since 1987. He studied first at the Central School then, like Daniel Barber, at Central Saint Martins, and this is his first job. He took his roughs – they were styling sheets rather than frame-by-frame storyboards – to Martin Foster immediately they had been approved at Channel 4.

> He came in with some lovely ideas – rough sketches – really early. We met and developed ways of bringing all five of them to life. When we started making the animations Joe kept coming in from time to time – steering us in the right direction. I did two, Ian Bird did one, and Mike Milne did two. The series won an Online Award in 1990.

They were all 100 per cent computer generated. No digital paint system was used. The very sculptured '4' in the version called 'Construction', which Martin worked on, was modelled on the 'Iris', as described above. From the wireframe the surfaces were rendered on 'Pixel' machines (Electric Image have two) which are ultra fast computers made by AT&T. The 'Iris' passes one instruction after another – admittedly a vast number of calculations – very fast but the 'Pixel' devices have 64 processors working in parallel. When they have instructions to perform they split them into 64 parts and each processor contributes to the whole task. They make even the most complex ray-tracing viable. On the 'Iris' each frame might take one to two hours to be rendered as a complex scene, on the 'Pixel' only 10 to 15 minutes! 'Pixels' only *render* – that is they convert the mass of digital data into on-screen colour, luminance and texture – no wireframe construction or motion, but they do that one job superbly.

This set of Channel 4 stings employed rendering, ray tracing cast shadows, reflections and refractions. Their style is not dominated by the medium of 3D computer animation and they demonstrate a state of harmony between high-technologists and a creative graphic designer. How much one relies on the other inconstant.

4 Getting a hand from the animator

Television designers have used the skills of the hand animator in their titles, special effects, and programme content throughout the past 60 years.

Some graphic designers have found ways of exploiting techniques relating to hand-drawn animation, for example rotoscope, or they have applied themselves to the rigours of hand-drawn cel work, with in-betweening, tracing and painting to achieve very high standards. A few people have left broadcast graphics to apply themselves solely to the animators' art. Most have called upon the skills of outside companies to bring to life their storyboards via the film rostrum camera, and more recently the video rostrum.

Hibbert Ralph Animation Limited is now a company that uses whatever form of moving-image production is most effective but it grew out of 'conventional cel animation', much of it for some of the best known television titles.

Jerry Hibbert says:

> Broadcast television graphics were very good at getting our business going and we made great progress with, and for, graphic designers throughout the 1980s. But as we developed, with more staff and better administration, we found ourselves out-growing the 30-second title made for as little as £10,000 and we moved from TV programmes to commercials.

> Only the simplest work could be achieved within the average television programme budget and this work gradually fell to the 'one-man-band' who could take on such work yet still make a profit. Some still do. Television graphic design simply could not afford us.

> Our work for commercials comes directly from the advertising agencies or the production companies, not from graphic designers. We work with them, or the art directors, once the production commences.

> Television graphic designers were always very demanding as clients – working with an agency we usually have more say over the final appearance of the work and we can have greater control over the style and pace of the animation we produce.

> The work we did with Pat Gavin for the LWT *South Bank Show* titles, over a number of years, always attracted favourable comment and captured awards due to the good working relationship between animators and designer.

Animation in commercials raised the designers' expectations of what could be produced but budgets were far higher – £90,000 for a single commercial is not uncommon – far more than broadcast TV could afford. We often had people coming to say 'We would like something like . . .' and then name a recent expensive commercial.

Among a mass of drawn animation for television commercials we have produced work for Toshiba, The Post Office, Addis, and Homepride.

The film rostrum is more or less superseded and we have turned over to video tape recording; we have a digital camera and we shoot on to an 'A60' laser disc, then we can manipulate artwork on Paintbox before it goes to 'Harry' for editing and sequencing. Film has gone. Now we are converting eight editing suites from film to digital video and laying soundtracks on to tape. Things are changing rapidly!

5 A computer program for drawn animation

The gap between what television wants and what it can afford to spend on 'cartoon animation' may be closed by a new system invented by Peter Florence and Andrew Berend of Cambridge Animation Systems.

A mere 30 seconds of drawn animation with, for example, five levels of activity could require 3,600 cels. The cost of in-betweening, tracing and hand-painting these has become so high that over the past few years a vast percentage of this work has been sent to the Far East and particularly to Manilla where labour costs are very low indeed.

The 1990 SIGGRAPH computer graphic convention was said to have 'revived' traditional animation, that is to say there were many films which mixed hand-drawn characters with 3D computer backgrounds. There were also

Cambridge Animation Systems have developed a programme that applies computer-control to the long pursuit of improving and speeding-up hand-drawn cel animation. This system, called 'Animo', has already been used in television title production by experienced designer/animators Pat Gavin and Hibbert Ralph, whose company has applied conventional hand animation to programme work and television commercials. Here are stills from Fence, *a commercial for The Post Office (Animators: Jerry Hibbert and Gethyn Davies/Camera: Peter Jones Rostrums) and another for Addis. (Animators: Kim Burdon, Graham Ralph and Alan Bassett/Rostrum Camera: Roy Lacey)*

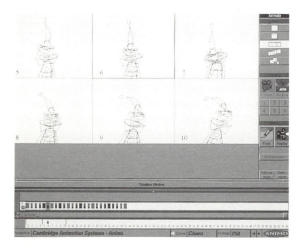

many new products developed to speed up traditional animation – an aim which had been tried in various ways in the previous 10 years. The Cambridge system is called Animo. It has been designed as a tool which is faster and cheaper than hand-work, yet one where the quality is high and the skill, as well as the style, of the individual animator is maintained in two areas – new methods of intelligent in-betweening and painting the frames. By using the system familiar to conventional cel preparation, the model of the character, or object, can be given red trousers, blue shirt, pink hat – the computer can paint and shade designated areas, then the whole animation.

There are several advantages. The line widths can be varied, but once set are completely maintained, or characters may be created without any outlines. Changes to colour and movement can be made very easily as work progresses. Coloured lines may be used. Shading – airbrushing – which is almost impossible to a high degree by hand becomes simple. Textures can be applied to any area by scanning artwork. Backgrounds, taken from any type of original, may be in put by camera and then panned at any speed and to any length.

When drawing subtle control of curves has been achieved by adding points, called 'beads', to the apexes at vital intervals on curves, allowing the software to manipulate the shape.

A great amount of time can be saved as line-tests are made and run on-screen as the in-betweens are created from one key drawing to another. Once an in-between has been made it can become the key frame for further movement or refinement. Sequences of character movement can be played over any background for the colours to be checked and everything previewed before the animator or director makes a commitment to record. A whole scene can be viewed and then the colour of any single element can be changed at hardly any cost. Multiplane levels are virtually limitless. Final output can be directly on to film via a standard film recorder, or on to video tape, or digitally to any resolution.

The Animo program incorporates an automatic lip-synchroniser which will recognise and select the correct mouth shape from a library of about 10 positions designated by the animator.

The system is being tested by animation studios in Europe and the sensitive and immensely large software package will run on standard hardware. Cambridge Animation Systems will guide customers on hardware and graphic designers in television may find hand-drawn animation will be accessible once again.

The revised opening sequence, designed by Pat Gavin, for LWT's *The South Bank Show* in autumn 1992 was the first TV title to employ Animo. Gavin reported that it animated a sequence of moving hands (drawn as a pastiche of Michelangelo) with 120 Animo frames from an input of only *five* key drawings. These used airbrushing and any attempt to animate in this way had not, in his experience been possible using hand prepared drawings or cels.

The voracious appetite of the television medium for moving images suggests the success of an advance like Animo will be of inestimable value.

6 New roles: management and the producer

Jayne Marshall is one of a more recent type of specialist in the world of television graphics. She learnt the business from 'inside' the most celebrated independent design groups – Lambie-Nairn and then McCallum Kennedy D'Auria. She gives this description of her function as a 'producer':

In the field of graphic design production there are obviously many ways in which any one job can be brought to the television screen. It could be conventional animation, computer graphics, motion control, studio or location filming, 'Harry' editing or just 'good old straightforward' VTR editing – the list of options goes on and on. However, several of these methods are adopted during the course of even the shortest sequence, whether title or TV commercial.

The producer in my role needs to have a sound knowledge of *all* these areas to provide accurate budget breakdowns to the client, to ensure that the job is completed within the budget agreed and also that no 'hiccups' occur, even on the most complex tasks. It is essential that every last detail is organised by the producer so that the designer/director can concentrate on the creative aspects.

At one time, when production was simpler, the graphic designer would struggle almost single-handed. Now, in the highly professional atmosphere of the independent design groups, the work involved in the liaison between client/agency and designer/director, the control of all costs and keeping everyone on schedule – often on more than one concurrent project – falls on the producer. Above all they must be 'jugglers' – par excellence.'

7 Animation with stills: film and video

Building up sequences of movements from stills can be done so skilfully that the viewers can be lulled into accepting that they have observed live action.

Shooting such material from prints, engravings, paintings, or specially prepared artwork was in the past assigned to the film rostrum camera. The working relationship between graphic designer and operator had been very close for many years. The trained operators' assessment of camera speed, lighting and exposures was essential.

Passages of information graphics are regularly required within news, documentary or arts feature programmes. This may be because no live-action film is available. In historical programmes, on periods prior to the late nineteenth century, the only source material will be still photographs, documents, prints or drawings, etc.

Stills sequences may need to be prepared with storyboards if they include many picture sources and are particularly complicated, to show the camera operator what is required. Otherwise the separate stills may be marked with field-sizes and the precise timings, on overlays. The pans or other movements required, the number of frames, and any fades and mixes (still the realm of film rather than video in this domain), must also be given to the camera operator. Stills sequences can form part of television commercials as well as programmes.

Ken Morse is outstanding among the exponents of stills animation. He learnt his craft at first when operating a rostrum camera for an animated film company before he went to the BBC in the 1960s. His name has appeared under the credit 'Rostrum Camera' so often he has become known internationally. He now works with both video and film from his own studio close to the many design groups in Soho and the West End. His sensibility and accuracy of timing are a legend, admired by directors and graphic designers.

At one time graphic designers *had* to rely on rostrum camera operators. With the advent of video cameras attached to rostrums, in almost every graphic design department and video workshop, the temptation to shoot 'do-it-yourself' stills sequences, or cobble something together using 'Harry', is very great. The lack of finesse of the carefully planned and exquisitely photographed sequences, of even the recent past, can be blatantly obvious.

8 Post-production and facility houses

The terms 'post-production' and 'facility house' are used in such a woolly way by everyone in the industry as to make them interchangeable.

Terry Hylton, the Creative Director of SVC Television in Wardour Street (another ex BBC trained graphic designer) suggested that *post-production companies* are best defined as those that have remained closest to the basic job of editing, while those who have assembled the wider range of digital video effects, electronic image-making and computer animation are better classified as *facility companies*. He elaborated as follows:

> Their roots are often in old film cutting rooms which adapted to meet the introduction of electronics in television production. Their early fundamental services – simple editing and the transfer of film to videotape – were spurs to growth, then they began to add captions and credits to titles and commercials. Once electronic equipment was invented the companies began to invest on a large scale in new 'hardware'. Nowadays the largest facility houses offer so much that they can be thought of as 'One-stop-shops'. SVC is high among these. You can go in with a storyboard, you can shoot graphic images, you can make complex graphic animation and post-produce or edit those elements – and walk out with the completed piece.

Such services are essential to graphic designers in programme and commercial work and Hylton describes what facilities, like SVC, provide. The majority are housed in the 'Golden Mile' of Soho.

> The 'big boys' bought their main 'toys' some while ago. Computers are now getting a lot cheaper and it is easier to get into the game than it was, at least for small companies who choose to work at a lower level, not the highest that is demanded by advertising agencies and broadcasters. The largest facilities companies are all very much dependent on commercials for the the bulk of their income.
>
> Appearing to work in the high-cost end of television commercials can be a 'turn-off' for some graphic designers but we have staff who are very capable in working to the shorter time-scales and the generally lower budgets of broadcast programme titles and effects. High spending is not a certainty,

even in the production of commercials in today's climate.

We are working more and more for our European neighbours and countries further afield. They are now aware that Britain offers production values which are higher than those expected or provided elsewhere.

(*Televisual* published a poll in 1991 listing facility companies throughout the whole of Europe rated as the best by their peers. This gave the top seven places to UK companies

Graphic designers in broadcast work are influenced by techniques and the quality of commercials, but on the other hand it is clear that some commercials have images which have been originated by graphic designers in television or in pop-videos. Sometimes it appears to be a circle. One instance is the technique of 'morphing' – seamlessly transforming one image into another – a special effect we were involved with as part of a Michael Jackson pop-promo. There then followed lots of commercials using the technique before it was very cleverly the basis for the Rory Bremner programme titles. The movement can go the other way and Bob English of English and Pockett did a curling typographic title for Channel 4 *Opinions* based on digital technology. This has been borrowed a great deal. Then there was a Maxwell House Coffee commercial which had 'steam/type' giving the message. Techniques and creative design influences travel in many directions (see Chapter 9).

To the questions 'How can graphic designers find out what each facility company does, and which ones should they use?' Terry Hylton replied:

All of the companies are constantly changing both equipment and personnel. I must admit it is very difficult for the graphic designers to keep up to date with the companies and all of the 'gizmos' they hold. What seems to happen is that the graphic designers build up relationships with the operators and editors. They then tend to stay with them unless the company falls behind in the technology stakes or, rarely, a project is handled badly.

To help our clients we hold open evenings with demonstrations to try to keep graphic designers and other users informed about new equipment, methods and costs and we are very proud of these. We also, as others do, provide showreels but they do not explain techniques and they are notoriously difficult to keep up-to-date themselves.

The technical press also plays a strong part in giving information about what we all do. Magazines have declined recently but *Televisual* and *Cuts* are both devoted to our area while Television Producer and *Creative Review* offer frequent well-illustrated features on television design and production.

Is the movement towards Apple Macintosh likely to come into this area of service for graphic animation? Hylton answered by saying SVC have not used it directly in production work so far but Lambie-Nairn & Company had asked SVC to compile a test animatic for a television commercial.

Their designer, Rob Kelly, was able to draw a large number of frames very fast on the Lambie-Nairn in-house Macintosh. At SVC we were then able to hook the Macintosh up to our system and get that information out on to a video format where we could do further work – adding colour and psuedo-animation to a music track. Between us we produced a highly finished result, combining the technology available to him in his studio with our own. Their client was so impressed with this animatic that they considered using it for transmission. This is definitely the way things are going.

Young designers coming from the colleges have a deeper knowledge of the computer systems than the previous generation and they are responsible for these developments.

SVC Televisions's range of 'kit' is the largest in the UK. For editing they have seven suites (one component digital, two component, and four composite) and a Harry. In video tape recording there are four D1, two D2, 26 one-inch, 18 BetaSP machines and two M2 machines. For electronic effects to manipulate images they have 13 separate channels linked to eight Abekas A53s, four Quantel 'Encores' and a 'Mirage'. In the area described as 'graphics' which means primary image making SVC operate two Quantel 'Paintboxes' – a Classic and a 'V' series, an EVS 'Paintbox', two 'Symbolics' paint systems with high-resolution 3D computer modelling and animation work and 10 character generators. Other facilities are three telecine suites and a studio with a motion-control rig.

An example showing the complexity of choice, that Terry Hylton acknowledges confronts graphic designers, is given in Chapter 7 section 13. □

Chapter 9   Inspirations and Influences

Very few artists or designers are secretive about their inspirations and influences. Most will discuss them openly and with enthusiasm. Awareness of other people's work is a part of every artist's mind. Nobody ever produces totally original ideas or style; nobody creates in a vacuum. The border-line between plagiarism and homage can be incredibly thin. Candour is best and sources should always be admitted.

A 'thesaurus' of images and wide general knowledge are a fundamental part of any designer's resource. While they will be attracted to particular artists, periods and styles, an interest in the wider visual world is imperative. Art and design history is not to be relegated to one day a week in the syllabus.

Influences are not contained by time or place. Guiseppe Arcimboldo, painter of heads composed of flora and vegetables from the sixteenth century, is just as likely to contribute to an animation exploiting 3D computer systems as is a painting by David Hockney.

The past at play

Sometimes the 'borrowing' is explicit and intended to be shared by the audience. In his design for the opening titles of the BBC2 series *Washes Whiter*, a study of the history of television commercials, Graham McCallum of McKD translated every detail of a collage, made by the artist Richard Hamilton for the exhibition. *This is Tomorrow* (1956), into a full-size set with astonishing accuracy. The original was less than 10 inches square.

Hamilton's early image of British Pop-art (the word 'Pop' appears reputedly for the first time as an emblem carried by the male figure) was entitled *Just What is it About Today's Home That's so Different, so Appealing?* and it portrayed social attitudes to advertising and consumerism. The paste-up of magazine cuttings was based on a typed list: 'Man Woman Food History Newspapers Cinema Domestic appliances Cars Space Comics TV Telephone Information'. A careful search reveals all these. The large size of the telephone in the

Constructing playful images from natural and man-made objects has always been a source of inventiveness to artists and designers and it still inspires much television animation. The work of Milanese artist Guiseppe Arcimboldo (1527–93) has sparked-off many of these. An arch-animator the Czech Jan Svankmayer (a still from Dimensions of a Dialogue *appears above) would certainly admit the influence of this fellow-artist, who also worked in Prague*

  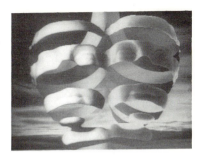

cuttings led Hamilton to put a woman telephoning on the television screen. McCallum and his art director, Raymonde Evangelista, who devised the set with him, mimicked even that.

McCallum's programme title uses this parody as a stepping-stone to the programme's content. He thought it was the first pop-art image to refer explicitly to advertising.

Although animations bring details of the collage to life throughout the 28 seconds of the title the complete picture is only on-screen for five seconds.

How many of the audience would have identified the original, shown over 30 years ago and not often reproduced since then? Those who did recognise it would have felt an affinity and enjoyed sharing the 'joke', and it was a perfectly appropriate way of establishing the period and introducing the subject to the whole audience.

Visual resonance can be detected everywhere. The whirling structures in a title made for *Forty Minutes* for the BBC were reminiscent of the playful kinetic sculptures of the Swiss artist Jean Tinguely (1925–1990). Was it based on his work? The inspiration may have been subconscious, or perhaps derived from other artists – Alexander Calder or Rowland Emmett?

Titles for Andrea Newman's drama series *A Sense of Guilt*, by the BBC graphic designer Iain Greenway, gave an instantly relevant sense of unease. The main character has the delusion that he can control the fate of those about him. Here the irrational perspective is based on *Another World*, one of a number of versions of this subject that appear in *The Graphic Work of M. C. Escher* (1898–1972). The concept is used appropriately and imaginatively – disturbing and intriguing the viewer. The viewpoints were made even more disorientating when animated.

Greenway realised his idea with a model (made by Alan Kemp) on the motion-control rig, using video, at

Gemini was a children's drama series produced by Thames Television and made at the Moving Picture Company. The central theme related to the mental relationship of twins and the elegantly executed combination of the coiling heads was hand-drawn cel animation in the pre-computer era

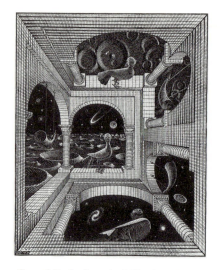
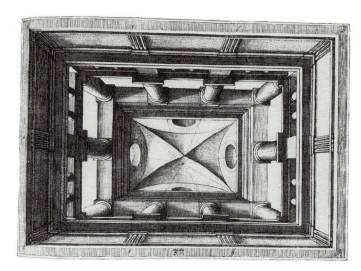

One of Escher's optical illusions clearly inspired Iain Greenway's animated title for A Sense of Guilt. *Could Escher himself have seen, and been influenced by, this 1600 engraving of a trompe-l'oeil design for a ceiling by fellow-Dutchman Vredeman de Vries?*

Cell Animation with Sebastian Witkin as the cameraman. Moving cloud backgrounds, combined with mattes in post-production, became the opening shot for each episode.

A French television programme on Escher included Greenway's compliment to him and part of the *A Sense of Guilt* title has already been further perpetuated in a pop-video!

Could Escher himself have been influenced by the fellow Dutch artist Hans Vredeman de Vries? The drawing here, from de Vries's book *Perspective*, was published in 1604. It too was made to establish an illusion. It was a *trompe-l'oeil* design for the ceiling of a loggia to give the observer the impression of greater room height.

Like other Surrealists, Escher's ambiguous treatment of space makes him particularly attractive to animators who can add apparent three-dimensional qualities to his brilliant optical illusions.

Another effective title, clearly based on an Escher wood engraving, was a children's drama series for Thames Television, *Gemini*, where the storyline featured twins.

The memorable graphic images of Magritte and Escher are in common currency and among the most frequently used. They only jar when their use is unsuitably strained or irrelevant.

There is pleasure for the audience in perceiving a reference to an earlier creator in both advertising and broadcast graphics, just as there is in recognising the source of a quotation in writing and speech. If the link with the past work helps to communicate then it is valid.

Head, heart and hand

Head, Heart, and Hand was the title of Professor Richard Guyatt's inaugural lecture when he was Head of the School of Graphic Design at the Royal College of Art. Implicit in his words is the relevant importance of imagination, personal inspiration and the level of craft work used in any creative process. All three elements must play some part – intellect, emotion and skill.

In painting and illustration there is a widely held predjudice in favour of the 'hands-on' concept. If a piece of work does not appear to be difficult to execute, or appears to be done quickly, then many people will dismiss it. Botticelli and Dali may be more readily admired because what they did was 'difficult'.

Jackson Pollock and Andy Warhol appear rejected because 'anyone can do it'.

The computer has aggravated this concept in television graphics and animation. Designers are often aware of the phrase 'It was all done on the computer'. This makes it all the more important to base design on a relevant idea.

The style of Georges Méliès early film tricks found their way into a Daily Mail *television commercial, made with the advanced technology of the mid-1980s*

From cinema to television . . .

Cinema has influenced television graphics. There are echoes, sometimes explicit pastiches, of the great film-makers – Méliès, D.W.Griffiths, Chaplin, Hitchcock et al, in many opening titles and commercials. A TV commercial for the *Daily Mail* used the animation style of Georges Méliès in a series of promotions for the newspaper. The images were borrowed from *A Trip to the Moon*, made by Méliès as early as 1902, perhaps his most famous film.

The animation style of UPA, formed by ex Disney animators who broke away in the late 1940s, was almost the backbone of television graphic animation in the black-and-white days of 1950s and 60s. Clones of John Hubley's 'Mr.Magoo' were plentiful.

Canadian animator George Dunning's *The Yellow Submarine* (1968), the post-war animation of the 'Zagreb School' in Yugoslavia, Bob Godfrey's independent spirit of *Kama Sutra rides again* and *The Wall* (for Pink Floyd), are all cinema sources having a strong influence on graphic design in television.

Jan Svankmayer's efforts to take stop-frame animation to the limits of the genre have also been flattered by sincere admirers, particularly in advertising animation. *The Snowman* and *Where the Wind Blows* animations, from Raymond Briggs' delightful books, have set a style for many television titles and commercials.

. . . and from television to cinema . . .

Recently the influence of cinema on television has been reversed where the look and technology of television was used by Peter Greenaway in his film *Prospero's Books*. The Quantel 'Paintbox' was used to conjure hundreds of richly illustrated pages, the imagined contents of the books mentioned briefly in Shakespeare's text. Greenaway's endorsement of the electronically-controlled 'Paintbox' is quoted in his book *Prospero's Books* (Chatto and Windus): 'The machine, as its name suggests, links the vocabulary of electronic picture-making with the traditions of the artist's pen, palette and brush, and like them permits a personal signature'.

Another influence on television graphic designers of all ages is the memory of 'classic' titles of the past. Years after their first transmission they can remain as a distillation of the whole programme

Thirty years after George Simenon's Inspector Maigret struck a match on the wall of a Parisian street (in dramatic black and white) to light his pipe, Geoffrey Martin's titles are still remembered with affection as an

requirements in conceptualization and operationalization for a meaningful relational analysis, item sampling adequacy, good item characteristics, the clarification of dimensionality of the items pool at different extraction levels, the articulation of various scaling assumptions guided by the CAUS, the systematic construction of multiple scales, the falsification or validation of the developed scales and scaling programs according to their measurement power or scaling effectiveness for the specific research purpose (emphasizing relational authenticity of the theory rather than absolute validity of individual constructs), and the multiple trials of selected scales in substantive analysis. Results from different studies can thus be compared against these universal criteria and standards.

Although many theoretical constructs, such as stress and social support, are fervently hailed for their significance to psychosocial studies, it is not well recognized that they have to suffer from being torn to pieces or reduced to some mere events in empirical research, or that they might be declared unscalable at all. It no longer helps to lament the myriad of particular ways of conceptualizing a field or the profusion of idiosyncratic measures. What is needed by the research community is to set up certain norms to help bring a study up to a new scientific standing. The key point here is the manner in which dimensions of a theoretical construct are reduced in order to facilitate its assessment. In other words, the central theme is unidimensionalization at various levels. In order for the reduction process to be scientific in the interest of appropriate generalization, some primitive requirements for unidimensionalized measurement at all levels could be explored.

First, any combination of measurement items must have some theoretical or empirical grounds showing their interrelationships as well as their relative importance to the resulting composite measure. It should further be based on some premises or facts about the relationships among all significant constructs involved in the study. The articulation of the theoretical assumptions or provision of empirical evidence in this regard as a norm will help the scientific community avoid confusion and many of the mistakes. Second, those assumptions and facts should have a solid theoretical foundation. This requirement helps to prevent unnecessary and counterproductive variations and inconsistencies in research results due to careless designs. It may appear to be too stringent, though, as theory in a field may be far behind. However, we should at least be aware that there are such assumptions underlying the operations, and the assumptions should have some rationale to show their plausibility in one way or another. In fact, these are the basic principles for scaling any complex theoretical construct, in which the

evocation of an era of television drama. *World at War* by John Stamp and Ian Kestle of Thames in the early 1970s, *The Old Grey Whistle Test* by Roger Ferrin for the BBC, the menacing snake in Dick Bailey's *I, Claudius* for the BBC adaptation of Robert Graves's novel, and bottle that still floats with a neon glare for *Arena*, never revealing its enigmatic message, was launched by Glen Carwithen when he designed the opening title for the BBC over 17 years ago; these and many others have attracted the sincerest form of flattery in dozens of subtle ways.

Perpetual animation

Image-making styles have constantly been recycled as one technical advance after another is born and then rejected. This can be witnessed in television graphics every few months rather than years. As soon as one new process is barely advanced, let alone 'perfected', there is a reaction and the next image-manipulating device is in place. The working lives of thousands of computer programers have been spent on complex algorithms to simulate natural phenomena in ray-tracing only to be spurned in a revival of free-hand animation – or so it seems.

Video feedback

The realm of video and electronic technology is a constant influence on the way pictures are presented.

The title for Ghosts in the Machine *collated 'a broad mix of video art and technologies' said Rick Markell of E&P when he made this award-winning animation for Channel Four in 1985. The programme gave genuine feedback. It was watched avidly by designers as a source of experimental video and film methods made by professionals and amateurs*

The graph of the oscilloscope screen, the coarse line formation of the television monitor, its basic pixellation and 'electrical-noise' have all been vehicles for image creation. The series *Ghosts in the Machine* (Channel 4, 1986) reviewed amateur and professional experiments in video. This was avidly studied by graphic designers and advertising agencies for new themes.

Pop-videos have been a rich, if sometimes over-indulgent, source of graphic images. The influences have been two-way, with broadcast and advertising television design both providing for and impersonating the genre.

Experiments, often financed on an enormous scale by universities in USA and Japan, and by military research, especially by the US Department of Defense, have over many years taken computer-graphic technology to a state of great refinement. Graphic imaging passed through various phases, using the computer influences of wire-frame models, the surface treatment of light, shade and colour – all those shiny reflective surfaces presented by the work of James Blinn and his collaborators throughout the 1970s – and then the mystery of fractals. All these are now an integral part of graphic presentation.

The winner of the 1991 annual Pentagram Prize for an outstanding essay on design, Barbara Wiedemann, a graphic design graduate of North Carolina State University, took as her theme the lack of historical background in design studies: 'We must acknowledge the fact that we cannot design without a cultural context.' There is, she concluded 'a responsibility to ask not only how great designers of the past created their work, but why'.

Designers in all mediums must have a highly developed awareness of the past as well as the moods and mores of their own time. □

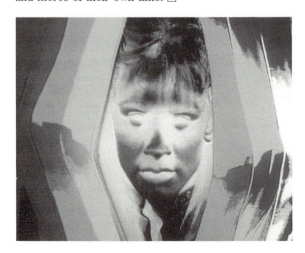

In the early application of new techniques there can always be a tendency to excessive use. The 'new toys' of computer construction led to styles, like wireframe and the highly-reflective surfaces of metal rendering, becoming epidemic. Although much good work was inspired by technology it took some years for the abundance of methods to be digested. The wireframe animation was for an Independent Television News title made in 1981 and the flash of animating metal (below) was designed by Lesley Friend, when she was a graphic designer at Channel Four News, just ten years ago

Glossary

Every profession and craft has its own jargon.

Television graphics has had to embrace all the technical terms associated with film animation and, since the entry of computer-aided graphic production, even more mystique.

The application of many of the terms is often disputed and they are not defined in standard dictionaries.

Many have been explained within the text and these may be traced through the index.

Aa

aerial image: a system where existing film images are projected on to a mirror set at 45 degrees below a rostrum camera unit. This enables animated sequences on cel to be combined very accurately with live action, frame-by-frame.

algorithm: in computer programing – a set of well defined mathematical rules for resolving a problem in a fixed number of operations. The word derives from the name of a ninth-century Persian mathematician.

analogue: 'the representation of values as a continuously variable signal'* as opposed to digital where the variation is carried out numerically. In short, using physical quantities.
(*From a SIGGRAPH paper.)

animatic: a very simplified animation made with rough artwork using only a few key frames. Usually recorded in low-cost video format as a demonstration for a client or director.

animation: any aspect of creating movement for film or television by providing a sequence of related images. The human eye's property of the retention of vision gives the illusion of a continuous image.

animation controller: electronic computer-based devices which allow drawings, photographs or other artwork to be VTR recorded, then viewed frame-by-frame, and edited in full colour.

animation cycle: short sequences of cels or video frames which can be repeated as required to lengthen an action, e.g. a figure running or walking, and reduce the workload.

animator: this term now applies equally to a hand animator or to someone working in computer animation.

answer print: after a film sequence has been edited, cut and graded the completed negative is printed. This, and subsequent prints, also called 'answer prints', are made for the client's approval.

anti-aliasing: the softening of the jagged edges, 'jaggies', (qv) which are an unwanted feature of curves and diagonals in raster display. Software adds pixels to the object's edges to create a blur. This technique is particularly welcome in the display of lettering.

archive material or footage: film or video material from a library source – this can be both historic and contemporary.

armature: a construction which allows models, usually covered with cloth or Plasticine, to be moved by tiny increments, for 'stop-frame animation' (qv) in both film and video.

aspect ratio: the proportion of the television screen's height to width – that is 3:4. Wide-screen experiments, at present using 'PALplus' as the transmission format, may result in a domestic picture of 9:16. A new challenge for graphic designers?

Bb

background: any painting or 3D model used behind moving figures or objects in animation.

back projection: moving film or transparency images projected on to a translucent screen. These are then recorded from the front – often with figures or objects moving in the foreground. Also referred to as 'BP' and by others as 'rear projection'.

barn door: metal shields on spotlights used to adjust the light beam.

BASIC: yet another acronym! A relatively user-friendly computer programming language, close to English – 'Beginner's All-purpose Symbolic Instruction Code'.

beads: points on a curve which can be picked up in computer animation and then manipulated.

binary: a numbering system having only two values – 0 and 1 where all other numbers are expressed with zeros and ones.

bit: contracted from 'binary digit'. The smallest unit of data in a computer usually represented as 0 or 1, a two-state code (binary). A single bit of information can be addressed by a computer's central processor at any time. The cluster of electronic points in the memory is known as the 'bit map'. Through a display controller the information in the bit map is converted into instructions to create the image on the video display.

bit map: see 'bit'

bug: a programming error causing fault in the computer.

byte: a store of eight bits which encode one symbol and produces 256 combinations. (eight cubed).

Cc

CAR: central apparatus room. The hub of technical operations in television transmission.

cathode ray tube (CTR): the video display device made of a vacuum glass tube where electrons are beamed to form luminous images on a fluorescent screen.

central processing unit (CPU): the computer function that interprets and executes instructions.

character generator (or 'cap-gen'): a computer which creates and composes lettering in the video medium.

chip: an integrated circuit, consisting of thousands of transistors on tiny pieces of a silicon chip.

cel: a thin sheet of clear acetate, originally celluloid (from which the term derives), carrying a series of images to create movement when filmed or video recorded – hence 'cel animation'.

colour bars: electronically-generated video colour-test system. (See back cover!)

colour-mapping: in computer parlance 'mapping' means applying. Hence 'texture mapping' and 'colour mapping' is using programs to apply patterns to any plane as well as three-dimensional objects. 'Bump mapping' allows patterns to be applied to the most uneven surfaces.

colour separation overlay (CSO): a video system allowing images from two or more separate television signals to be viewed separately but presented as one picture. A key colour is required and this is most often a bright blue.

composites/compositing: combining multiple images, still or moving, within the camera was a major preoccupation in the film industry from its inception. Many elements can be used to make up a single shot, and video and electronic means have merely increased the facility and complexity of these 'tricks'.

computer-generated animation: images first created within the computer, purely as digital coding, are later displayed on a CTR screen by-passing drawings on paper.

conversion: film can be converted to video and video to film. There are many rival systems. Conversions are often needed from one international television format to another – e.g. NTSC to PAL, PAL to SECAM, etc.

coordinates: a set of data which specify a location in a model image or device space.

corporate: design or production work for any non-broadcast use.

crawl: lettering, no matter how it is produced, which moves into screen horizontally, from left or right.

cursor: a small mark on the monitor – a dash or a cross – to indicate the active screen position of the instruction.

cut-and-paste: digital paint systems have been designed to mimic all the actions of conventional artwork. The use of the words 'brush', 'pen' and 'palette' shows the old vocabulary has been copied as well.

cut-off: the area of a graphic or other image lost when a picture is transmitted by television.

cutting copy: in film editing all the chosen sequences are put together as the 'cutting copy' before being sent to the processing laboratory and matched to the master negative.

Dd

data: the singular of the Latin *datum* (give) meaning 'a thing known or assumed'.

digital: representing values by numbers.

digital effects devices: all manner of electronic equipment has been produced to manipulate images in the video medium. Many are listed in Chapter 6.

digital paint system (DPS): the ability to 'draw and paint' directly into the video medium was the most direct contribution by electronics to the elevation of graphic design in television. Quantel's tradename 'Paintbox' has become generic.

digital picture library (DPL): an electronic stills store containing limitless images as frames of information. These are instantly accessible to graphic designers when building television programmes now vital to news transmissions. (See 'electronic stills store'.)

digitiser: an electronic device to code images into numerical information by capturing 'x' and 'y' coordinates at any required interval.

digitising: transforming pictorial information of any kind by numerical data into the computer memory. This can be done by hand, by plotting points with a stylus, a mouse, or through a digitiser. Typically a wireframe object is constructed in this way.

digitising tablet: see 'tablet'.

disc: a storage device for electronic data, qv. 'hard disc' and 'floppy disc'.

disc drive: this mechanism not only revolves the disc, it 'reads and writes' the information.

dolly: a movable 'truck' designed to support a movie-camera on wheels as steadily as possible.

dope sheet: a detailed instruction sheet prepared for a rostrum camera operator by the graphic designer or animator. Every frame, mix, fade and multiple camera exposure and movement is described to synchronise with the soundtrack. With instant playback in video these are becoming less common. ('Dope' US slang for information.)

downtime: time when the computer is unused. Now extended to occasions when a whole facility, equipment and staff may not be busy.

dubbing: this is the process of copying an audio or video sequence or a programme. Dubbing is also superimposing an alternative voice over the original. (i.e. a foreign language or a trained voice over that of a non-singer).

DVE – digital video effects: all those page-turns and wild picture movements that graphic designers often regret as they are applied without any design consideration.

Ee

edit suite: rooms banked with video tape machines. These have largely replaced film editing and cutting rooms in the graphic designer's production process.

editing: the process of compiling sequences in film or video to within a single frame in either medium. The term 'cut' is common to both systems. 'Time coding' (qv) allows the video recorder to locate any individual frame and its number is displayed on the monitor.

effects device: see 'digital effects device'.

electronic library: when the capacity of stills stores extends to a vast permanent storehouse of images – accessible to all the programme-makers – it becomes a 'library'.

electronic stills store (ESS): this invention allows graphic designers to hold and retrieve the images they create in the electronic realm. It was the key to the development of electronic graphics. The days of paper, paint and glue were over!

endoscopic lens: a lens system which allows filming to be carried out inside an object or confined space. (*Endon* = Greek 'within').

ENG: electronic news gathering.

Ff

facility house: a company providing image-making and post-production services for all types of film and video animation. Synonymous with 'post-production house'.

field: a single frame of television image made from two alternate 'fields' of information in the PAL system.

field size: (nothing to do with the previous term!) this describes the area of the artwork being recorded in film or video on the rostrum bed. The dimension given is the horizontal (i.e. 'a 12-inch field size') and the aspect ratio will be selected film or television. Sets of field charts can be obtained as clear acetate templates to be laid over artwork and these define the inner 'safe area' transmitted in television.

film recorder: a device for transferring digital-based data from video on to film.

floppy disk: a small flexible disk which stores video images.

footage: a prescribed length of motion-picture film.

Fortran: this is a high-level computer programming language – the term taken from '**For**mula **Trans**lator'.

fount (or font) compose: any device which allows the operator of a character generator to build new typefaces from any design or source into the electronic system.

fractals: a computer programming term which relates to the modelling of complex irregular forms (e.g. water, rock formations or clouds). Chains of repetitive calculations resulting in random rather than regular geometry. Developed in 1975 by Benoit Mandelbrot, a French mathematician. Fractals aid computer programmers to create the 'chancy intricacies of nature'.

frame: a single image in film or video. Normal film speed for recording and projection is 24 frames per second. A video frame in PAL is made up of two fields of alternate lines and there are 25 frames per second in that format.

frame buffer: an area of a computer which stores image data temporarily while being processed within a computer system. 'A holding area for pixel data which is to be displayed.'

frame grabbing: any number of still images – single frames – can be selected and recorded from a main frame store. Still images are said to be 'frozen'.

freeze frame: a single frame fixed on-screen from any video source.

Gg

gallery: the control centre for a television studio. They contain monitors showing all camera and other video inputs, electronic switchers and DVE machines. When electronic graphics were in their infancy graphic designers spent much time in these areas. Now post-production work takes place in fully equipped electronic graphic workshops.

glass shot/glass painting: by painting part of a scene on a large sheet of glass a set or a location can be extended. Filming through the glass, with a real scene in the foreground, could achieve an otherwise impossible effect or save money on the construction costs (e.g. the upper storeys of a very elaborate building). Electronic techniques, colour separation overlay among them, have virtually by-passed this need.

grading/graded print: part of colour-film processing where the aim is to make the final print as consistent as possible from shot to shot.

graphic workshop/electronic graphic workshop: any collection of devices (hardware and software) allowing the production of on-screen images or lettering by electronics.

graphics system: see 'graphic workshop' etc.

Hh

hard copy: a permanent copy of an image taken from a CRT – this can be a colour print or transparency.

hard disc: these are large capacity stores built into electronic hardware.

hardware: the physical apparatus of computer graphic and electronic image-making – circuit boards and peripherals (cf. the programs, documents and routines used to instruct computers which is the 'software'). Very high investment is usually involved.

hidden line: the term used to describe the removal of those lines from a computer wireframe construction which converts that image to one where it appears to be solid, rather than transparent.

high-speed filming: some cameras are designed to shoot at far more than 24 frames per second and they can capture split-second action in a very graphic way. This is a technique where film is more effective than video.

howl: this describes the video effect where the outline of an image is 'echoed' many times almost to infinity.

Ii

in-between: a drawing, or computer construction, of any figure, or object – made while preparing an animated sequence – to link two key positions.

in-house: a service, facility or staff resource owned by a company or organisation of any kind.

input camera: these are video cameras, usually mounted on a simple rostrum, where designers working in an electronic graphic workshop are able to record any piece of artwork or photograph in a few seconds.

insert: replacing a sequence of film or video material in previously recorded material.

interface: a boundary between any two systems in the graphic production process.

interactive: the idea of a dialogue between the operator and an electronic device at very high speed (see 'real-time'). 'Interactive video' also describes the use of the medium for teaching purposes.

Jj

jaggies: the crude staircasing effect on diagonals in raster display. (See anti-aliasing.)

Kk

keyline: this is an outline defining any area which may then be separated from its surroundings. Once filled as a solid it becomes a matte or master.

kilobyte: a measure of the capacity of computer memory (1,024 bytes equals one K, i.e. 2 to the power 10).

Ll

light source: in 3D computer graphics the models created can be shaded to achieve realistic effects as if lit from any position and with any light intensity.

line-test: as multiple drawings are produced in animation it is essential to review them frequently. Any means of doing this is called a 'line-test'.

live-action: filming or video recording any real event.

locked-off camera: a film or video camera that is used in one absolutely fixed position throughout the shoot.

logotype/logo: originally a metal casting of a single word. Now the term used for the design of the title of a programme as a 'trademark'.

Mm

master: see 'mattes'. A master is also the completed version of an animated sequence or TV programme from which copies can be produced.

mattes: stencils or masks. In film these were painted on cel with opaque material. In electronic devices they are created with the cordless stylus or mouse. They can be positive or negative shapes to a 'master'.

menus: to the television graphic designer these are the lists of forthcoming programmes which they designed for many years as single caption cards or slides and now prepare on digital paint systems or sometimes animate. In computers they are the list of options displayed on a monitor allowing the operator to choose the next action to be carried out.

mix: originally the American term for fading from one scene or sequence to another. Now widely superseding the word 'dissolve'.

mnemonic: the word adopted by television graphic designers to describe the extremely short animations – only two or three seconds – of tradenames, symbols or station idents they create as a way of making them more memorable when presented on-screen (e.g. the opening umbrella for Abbey National or the swiftly-formed classic head for Prudential).

modelling: in three-dimensional computer graphics this is the mathematical rendering of an object which can then be moved and viewed from any position.

monitor: the same as a cathode ray tube – CRT.

morphing: the computer's ability to manipulate many subtle in-betweens connecting any two key drawings or photographs – recently highlighted in changing one famous face into another.

motion-blurr: programs have been written to reduce the over-sharp images associated with computer animation. Motion-blurr gives greater reality.

motion-rig: a unit where a film or video camera is fixed to a structure where it can be moved to any position in a studio. Camera operation, the rig, and sometimes the armature holding a model, are nearly all computer-controlled for accuracy and possible repetition.

movieola: a device where rush prints (qv) can be viewed and heard and film may be synchronised to sound. Also known as a 'pic-sync'.

multiplane: artwork under a rostrum camera can be set up with more than one level of glass backgrounds and foregrounds to give a greater sense of depth to the animated picture. They may be lit separately.

mute/mute print: video tape or film print with picture only – no sound.

Nn

nanosecond: to show how fast computers can operate – one-billionth of a second.

nested zoom: a camera movement into or out of a section of the picture area – not the whole frame. Used frequently in black-and-white 'B' movies.

noise: on-screen disturbance caused by electronic interference.

Oo

OB: outside broadcast equipment or personnel.

offline: work carried out, or hardware used, outside the direct transmission process.

on-line: hardware directly linked to the transmission process or to the central processor in a computer system, as opposed to a device being described as 'off- line' or 'stand-alone' or a final edit prior to transmission.

oscilloscope: a device which displays on a CRT, in waveform, the performance of electronic equipment so that optimum results may be obtained.

Pp

paint system: a digital device which allows graphic designers to 'paint and draw' directly into the video medium using a stylus and digitising tablet. Its arrival was one of the most revolutionary steps in graphic design for television.

pan: vertical or horizontal movement with a rostrum camera bench over artwork. Similar moves with a live-action camera on a scene. Hence 'pan-up', 'pan-down' and 'pan-left' or 'pan-right'. Sometimes the terms 'north', 'south', 'east' or 'west' are used.

Pascal: one of many computer programming languages with a precise format of definitions, procedures and data.

Picaso (sic): a computer graphic system developed by Dr John Vince which used routines from the high-level computer Fortran.

pic-sync: a machine used in film animation to link the separate frame images to a soundtrack.

pitch: a word borrowed from advertising now that television graphic designers are more likely to be competing for projects.

pixel: this is the smallest on-screen element in the computer display system. The word is made from 'picture element'. A full-colour television frame contains 400,000 pixels and each of these requires a digit to represent the colour signal red, green or blue. So a mere 1,200,000 numbers must be addressed every twenty-fifth of a second to display a full-colour animation!

plotter: a device controlled by a computer to draw images on to paper or film with precision and at high speeds.

post production: the processes carried out *after* the creation of the original sequence of images. These may be to combine with others or to enhance them in any way.

pre-roll: the space of time required for telecine or VTR equipment to provide stable pictures.

program: information to control the operation of a computer. Not 'programme', which describes the material transmitted on television.

Rr

random access memory (RAM): data stored in computer hardware that can be 'written to or read from' in any order.

raster: the system that uses the hundreds of horizontal lines of phosphor elements (qv 'pixels') on the video screen. Hence 'raster graphics'.

ray tracing: programming computers to imitate the natural behaviour of light – how it travels, how it is reflected, how it is absorbed – all these and many other properties are reproduced by extremely complex mathematics. Ray tracing software is the main rendering technique to advance the appearance of images in television graphics.

read: the central processing unit in a computer is said to 'read' when information is examined and transferred to its memory.

real-time/real-time animation: an animation carried out by a computer program where there is no perceivable delay between the instruction and the reaction. Computers must have huge processing power to display three-dimensional, coloured objects on-screen with the appearance of instantaneous movement at 25 frames per second.

rendering: in computer graphics there are two main elements – modelling and rendering. Rendering describes the model's highlights, shadows, colours and textures.

resolution: the quantity of pixels available to display the on-screen image, e.g. small number of pixels = low resolution.

RGB: the red, green and blue phosphors arranged in vertical columns and horizontal rows on a CRT that emit coloured light when they are struck by colourless electron beams.

roll/roller: both terms originated when end credits for programmes were set with metal type and printed on long rolls of black paper. Both terms have transferred to electronic typesetting.

rostrum camera: a film or video camera mounted above a bench on a tall column to record animation. Both bench and camera are engineered to allow extremely small and precise movements both up and down and to north, south, east and west.

rotoscope: any device which enables rotoscoping to be carried out. A rostrum camera or specially constructed equipment can be used. See previous entry.

rotoscoping: this was a film-based process where film images, usually human or animal figures, were projected on to a ground glass surface and then traced frame-by-frame to aid or create unusual animation. It can now be achieved with a 'Harry'/'Paintbox' configuration.

rundown: an on-screen list of forthcoming programmes.

rush print/rushes: these are the first, ungraded film prints sent to the production company. The graphic designer and others can then decide if a re-shoot is required before proceeding.

Ss

set-props: graphic designers have always been responsible for designing and producing all the printed ephemera such as fake banknotes to signs for a fictitious railway station in eastern Europe.

showreel: a collection of work representing the output of an individual designer or group. The old film term has carried over to any video format.

slave: a VTR or other machine controlled by a second machine.

slit-scan: the extreme precision given to the computerised film rostrum camera which was developed by this technique of photographing artwork in motion through a narrow slit. With careful control and much patience, artwork could be distorted and perspective given to two-dimensional images. Digital video effects devices have more recently achieved similar manipulations electronically.

SMPTE: the Society of Motion Picture and Television Engineers (USA) that sets many of the standards that are accepted internationally.

snorkel lens: a lens system using prisms and mirrors to enable movie cameras to record images in otherwise inaccessible positions.

software: these are the programs and instructions given to computer hardware to make them work, either 'made- to-measure' to the graphic designer's purpose or a 'software package' from a manufacturer. e.g 'SoftImage' or 'Wavefront'.

special effects: this term is sometimes used to describe the skills needed in physical ways – fire, water, explosions – sometimes it defines electronic and photographic techniques in filming, video recording and in post-production.

split-screen: in film or video there are many ways of presenting two, or many more, independent moving images.

sting: the briefest animation to announce any forthcoming television series or individual programme.

stop-frame animation: both film and video can shoot one frame at time to build-up limitless speeds and effects in animation.

storyboard: a plan of action, which can be visualised in any style thought suitable to convey the graphic designer's intentions to the director or client.

streak-timing: the computer-controlled rostrum camera, when using film, can retain the streaks of light that appear when long exposures are used. Back-lit negatives used with coloured gels could trace quite complex patterns.

Tt

tablet: the electronic pad in a digital paint system which connects the movement of a 'pen' or 'mouse' to to the on-screen cursor through a grid of hidden wires by encoding 'x-y' coordinates.

telecine/telecine machine (TK): the processes concerned with transferring and transmitting moving film or colour transparencies to the television medium are all referred to as 'telecine', e.g. a telecine slide scanner allows the input of any 35 mm slide to the raster display of a digital paint system.

terminal: a point at which a keyboard, stylus or mouse is used by a designer/operator produces images or lettering on the screen.

texture mapping: computer-generated images made more realistic by applying a two-dimensional pattern or texture to the polygonal three-dimensional models. The edges are smoothed by anti-aliasing.

time code: a code number is recorded on every frame as an aid to synchronising and editing video tape.

timelapse: by exposing a single-frame movie camera at very delayed stages in a process of movement the final projected film can speed up any action with very graphic effects. Changing skies or a growing plant are classic examples. Similar results could be obtained by film and video editing.

three-D animation: the immense problems of depicting perspective from a moving viewpoint confined most hand drawn animation to parallel perspective with a centre of vision and two-dimensional figures. Computer-aided animation made the exploration of space more practical.

topping and tailing: the word 'merely' usually precedes this expression which implies the creative design effort and production costs of making the opening title – 'topping' – and preparing the end credits – 'tailing' – are to be absolutely minimal.

total costing: in the past the production of graphic work was only quantified in terms of materials and services incurred outside the television contractor's resources. Now the costs of all those contributing are accounted – graphic designer's hours, assistant's time etc.

travelling mattes: in film-work these are masks on high-contrast film which move from frame to frame to achieve composite shots (e.g. a live-action figure sequence combined with a background shot at a different time and place).

Uu

U-matic: the three-quarter-inch video tape format produced by the Japanese company Sony.

Ultimatte: a patent system for colour separation in video which allows two or more image sources to appear in a single frame.

Vv

vector: a line drawn between two points in two or three dimensions. Also the system presenting lines, as opposed to raster (pixel) display, on a monitor screen.

VHS: the half-inch video tape format produced by Matsushita of Japan.

methodological crux is how to reduce the complexity of data by unifying different dimensions represented by various items and subscales. In this regard, the CAUS attempts to reestablish the integrity of important theoretical products by making them scalable "as-is" (at the generalized levels). It may be used to develop scales using specific data sets to tackle the technical issues in measurement, but is aimed more fundamentally at illuminating the way in which psychosocial scaling could be done more orderly, reasonably, systematically, and effectively.

In a sense, a scale is a most efficient device of data reduction with supposedly minimum loss of information. The idea of unidimensionalization is the key to multidimensional measurement and analysis. The issue that the CAUS tackles is how to synthesize the results of dimensional analysis, and obtain a variable that can be directly measured and represent the original theoretical construct. The CAUS does not only systematically summarize the current practice in social science research, but also lay a solid groundwork for resolving the difficult problem through comparison with natural sciences.

Validity, confounding, and the understanding of purpose

Validity is a (if not *the*) most important and ingrained concept in measurement and scaling theory. Although the term is somehow overused to imply "all things good about a measuring instrument" (Nunnally, 1978, p.86), psychometricians do attempt to make it explicitly indicate "the standards by which measuring instruments must be judged" (ibid.). The special effort they have made is to distinguish between the different meanings of the idea, which has resulted in different types of validity employed in scaling practice. Preoccupied with "what it is supposed to measure" in terms of a certain universe of content (specifically a psychological attribute), psychometricians have focused on scaling individual constructs. Their embeddedness in validity has been accompanied by an attention mainly paid to unidimensional constructs plus a consideration of the method factor and response pattern or some general measurement conditions. To maintain the underlying assumptions and validate various scales, analytic procedures such as confirmatory factor analysis (CFA) are valued (McArdle, 1996).

Nevertheless, a "chronic crisis" in psychological measurement and assessment is observed in terms of the problems in validation (Meier, 1994). Meier (ibid.) attributes the problems to a considerable ambiguity in the application of

video cassette recording (vcr): the rapid developments from the late 1950s in recording images on to tape and editing them electronically have been fundamental to progress in graphic design work in television.

Ww

whip pan: a very rapid movement of a camera from one point of a scene or artwork to another.

wire frame: the now familiar construction of straight lines used in computer animation to display any figure, scene or object. The system was anticipated in the fifteenth century by Peiro della Francesca and Uccello in their perspective studies.

work-station: electronic graphic equipment grouped at one point to allow a graphic designer/operator to produce a wide range of on-screen material. A paint system and character generator are the basic hardware.

write: if computers can 'read' they can also 'write' – which means they record information in their memories.

Zz

zoom: a movement obtained by the camera or by a lens to take the observer closer or further away from the subject – hence to 'zoom in' or 'zoom out' – usually relatively fast. At slow speeds the terms 'tracking-in' and 'tracking-out' are normally applied.

Addresses

ADAR
Art and Design Admissions
Registry
Penn House
9 Broad Street
Hereford HR4 9AP
Telephone: (0432) 266653
Applicants with disabilities
contact 'Skill'
National Bureau for Students
with Disabilities
336 Brixton Road
London SW9 7AA

BAFTA
British Academy of Film
and Television Arts
195 Piccadilly
London W1V 9LG
Telephone: (071) 734 0022

BTEC
Business and Technology
Education Council
Central House
Upper Woburn Place
London WC1H 0HH
Telephone: (071) 388 3288

BECTU
Broadcasting, Entertainment,
Cinematograph and Theatre Union
(previously ACTT and BETA)
111 Wardour Street
London W1V 4AY
Telephone: (071) 437 8506

BKSTS
British Kinematograph Sound
and Television Society
M6–14 Victoria House
Vernon Place
London WC1B 4DF
Telephone: (071) 242 8400

CSD
The Chartered Society of Designers
29 Bedford Square
London WC1B 3EG
Telephone: (071) 631 1510

D&AD
Nash House
12 Carlton House Terrace
London SW1Y 5AH
Telephone: (071) 839 2964

RTS
Royal Television Society
Holborn Hall
100 Grays Inn Road
London WC1X 8AL
Telephone: (071) 430 1000

Magazines
Broadcast
EMAP
33–39 Bowling Green Lane
London EC1R 0DA
Telephone: (071) 837 1212

Creative Review
Centaur Limited
St Giles House
50 Poland Street
London W1V 4AX
Telephone: (071) 439 4222

Television Week
EMAP
33–39 Bowling Green Lane
London EC1R 0DA
Telephone: (071) 837 1212

Televisual
Centaur Limited
St Giles House
50 Poland Street
London W1VB 4AX
Telephone: (071) 439 4222

TV World
EMAP
33–39 Bowling Green Lane
London EC1R 0DA
Telephone: (071) 837 1212

Exhibitions
Parigraph
IMV/Parigraph
17 Rue Lamandé
75017 Paris
Telephone: (33/1) 42 29 88 00

Computer Graphics
Blenheim Online
Blenheim House
Ash Hill Drive
Pinner
Middlesex HA5 2AE
Telephone: (081) 868 4466

Television Broadcasters
BBC Television
Graphic Design Department
Scenic Block
BBC Television Centre
Wood Lane
London W12 7RJ
Telephone: (081) 743 8000

Channel 3   The ITC companies:
(Formerly the IBA)

Independent Television Commission
(The public body responsible for
licensing and regulating non-BBC
television services)
33 Foley Street
London W1P 7LB
Telephone: (071) 255 3000

Anglia Television Limited
Anglia House
Norwich NR1 1JG
Telephone: (0603) 615151

Border Television PLC
The Television Centre
Carlisle CA1 3NT
Telephone: (0228) 25101

Central Independent Television PLC
Central House
Broad Street
Birmingham B1 2JP
Telephone: (021) 643 9898

Channel Television Limited
Television Centre
St. Helier
Jersey
Channel Islands JE2 3ZD
Telephone: (0534) 68999

Carlton Television Limited
101 St Martin's Lane
London WC2N 4AZ
Telephone: (071) 240 4000

GMTV Limited
The London Television Centre
Upper Ground
London SE1 9LT
Telephone: (071) 827 7000

Grampian Television PLC
Queen's Cross
Aberdeen AB9 2XJ
Telephone: (0224) 646464

Granada Television Limited
Granada Television Centre
Manchester M60 9EA
Telephone: (061) 832 7211

HTV Cymru Wales
The Television Centre
Culverhouse Cross
Cardiff CF5 6XJ
Telephone: (0222) 590590

HTV West
The Television Centre
Bath Road
Bristol BS4 3HG
Telephone: (0272) 778366

London Weekend Television Holdings PLC (LWT)
The Television Centre
Upper Ground
London SE1 9LT
Telephone: (071) 620 1620

Meridian Broadcasting Limited
Television Centre
Southampton SO9 5HZ
Telephone: (0703) 222555

Scottish Television PLC
Cowcaddens
Glasgow G2 3PR
Telephone: (041) 332 9999

Tyne Tees Television Limited
The Television Centre
City Road
Newcastle upon Tyne NE1 2AL
Telephone: (091) 261 0181

Ulster Television PLC
Havelock House
Ormeau Road
Belfast BT7 1EB
Telephone: (0223) 328122

Westcountry Television Limited
Western Wood Way
Langage Science Park
Plymouth PL7 5BG
Telephone: (0752) 333333

Yorkshire-Tyne Tees
Television Holdings PLC
The Television Centre
Leeds LS3 1JS
Telephone: (0532) 438283

Satellite broadcaster
BSkyB
(British Sky Broadcasting)
6 Centaurs Business Park
Grant Way
Isleworth
Middlesex TW7 5QD
Telephone: (071) 782 3000

Television news
Independent Television News (ITN)
200 Grays Inn Road
London WC1X 8XZ
Telephone: (071) 833 3000

Manufacturers
Aston Electronic Designs Limited
123–127 Deepcut Bridge Road
Deepcut
Camberley
Surrey GU16 6SD
Telephone: (0252) 836221

Quantel
Turnpike Road
Newbury
Berkshire RG13 2NE
Telephone: (0635) 48222

End credits

What a ponderous machine a book is to put together,
more important, what a complicated one!
  Gustave Flaubert/Letter, 1852

My recent discovery of Flaubert's warning came far too late, but throughout the long preparation of this book I have gained great pleasure through the generosity of many people who have given me that most valuable commodity – their time. By visiting past colleagues, as well as making many new contacts, I have been able to investigate what they continue to do with so much skill and enthusiasm.

My thanks are given to: Brian Tregidden, Head of Graphic Design/BBC for guiding all my requests to the designers in his department for over a year; Peter Le Page, Controller of Design at Thames Television, for his loyalty in the past and recent advice; Margaret Riley of Focal Press for commissioning and encouragement.

To all those who put so much effort into making Chapter 7, 'The Reel World' a reality through interviews, written contributions and many illustrations, my very grateful thanks: Morgan Almeida, Daniel Barber, Mike Bennion, Liz Friedman, Richard Higgs, David James, Chris Jennings, Martin Lambie-Nairn, Graham McCallum and his colleagues at MCKD, Barry O'Riordan, Marc Ortmans and his partner Haydon Young, Jean Penders, Morgan Sendall, and Kasuo Sasaki who I met by happy chance at Middlesex University.

For other design contributions my gratitude to: Bob English and Darrell Pockett of E&P for their 'video-type' designs, Iain Greenway of BBC Graphics, Al Morrison and design students of Westminster University, Peter Pegrum, Design Manager/Visual Effects/BBC, Joe Roman for his Channel 4 animations.

For the interviews about the services that support graphic production my very warm thanks to: Paul Doherty of Electric Image who has always found time, Doug Foster of Cell Animation, Martin Foster at Electric Image, Peter Florence and Andrew Berend of Cambridge Animation Systems, Jerry Hibbert and Anna Hart of Hibbert Ralph Animation Limited, John Holton of Aston Electronic Designs, Terry Hylton the Creative Director at SVC Television, Alan Kemp the model-maker, Sue Lipscombe of Codsteaks, Jayne Marshall of MCKD for her description of her role, Ken Morse (who deserves all those screen credits) and Robert Pank at Quantel, who was of great assistance and who enabled Paul Rigg to contribute information and illustrations.

Artwork for the cover design was prepared by Paul Rigg, through the generosity of Quantel Limited, on a Quantel Graphic Paintbox which has a 14,000 line resolution. Acknowledgements are due to Lambie-Nairn & Company for their kind permission to reproduce their design for the BBC2 ident in this instance.

My thanks for other contributions also go to: graphic designer Dick Bailey of the BBC; Derek Carr, ex-HTV and now researching at Bristol University, who took such care in his advice on electronic engineering that he gave new meaning to the word PAL, Colin Cheesman, whose contributions on design education and his own career were invaluable; Hazel Dormer of Channel 4 News; Peter Ming Wong and Kieran Walsh, who took time to give accounts of their early explorations in graphic design; Peregrine Gibbs for his excellent dissertation and videotape; Paul Conduit-Griffin, who stimulated my thoughts by his researches in graphic technology; Mike Hurst, Head of Graphic Design at BSkyB and Ted Simmons, Head of Presentation at Carlton Television, who gave valued information about their new worlds; Alan Jeapes of the BBC, who allowed me to quote from his Monotype lecture 'Sex, Type and Videotape' and who was very good company. My thanks to Mitch Mitchell of the Moving Picture Company for his quoted comments on training and T. Dilwyn Morgan MD of EOS for animation control information.

Excerpts from Norman McLaren's Technical Notes, and illustrations of his work, are reproduced by kind permission of the National Film Board of Canada.

Further advice and very thoughtful contributions on electronic engineering were made by Tony Watson, of Techinical Television Training, who is a Fellow of the Royal Television Society.

Expert advice on the use of my PC was always generously bestowed by Ray Smith of Computer Support, Bristol.

Janet Margrie and Anna Rose read the text with great care and concern to minimise errors. Those that remain are the fault of my own random access memory.

Don Shaul of Middlesex University gave much advice.

DM/Bristol 1993

Index

Campbell and Fiske's (1959) criteria to demonstrate convergent and discriminant validity. The real issue, however, is with the limitation of the notion of validity itself. The conventional idea of validity focuses on individual constructs yet overlooks the relationship among the constructs as well as the necessary conditions to validate such relations. In real terms, the idea of construct validity even tends to deny or preclude any significant relationship among different constructs, since if such relationship does exist, it would easily be interpreted as an indication of low discriminant validity in measurement. On the other hand, the achievement of convergent, concurrent, content or any other kind of validity does not necessarily mean a good measuring instrument as required by a relational analysis.

In real terms, a major problem can be identified in empirical research, that is, lack of a guideline for conceptualization and operationalization. Many research commentators have cited the issue of theoretical confounding in relational studies. No one, however, has articulated downright a way out, while most research workers are not even certain whether or not this problem is avoidable. Indeed, psychometric studies of validity contributes little to theory development with regard to the increasing appeal to better measurement tools as a solution to theoretical confounding in empirical research. Not only is conventional validity unable to resolve the confounding problem, but the pursuit of validity is more likely to contribute to it, especially when people pursue validity without purpose, or try to establish universal or all-purpose validity. In this regard, the distinctive logic of a relational analysis must be stressed, which will help redress the role of validity in scale development and address the paradoxical needs for both conceptually valid and theoretically useful measurement.

Judging whether a scale could be a good one in a relational analysis needs to take into consideration the conceptual boundaries between different constructs. Beyond conventional understanding of the issue of confounding, a dilemma underlying all conceptual boundaries should be recognized by relating to the practice of questionnaire design, in which reversing the coding of some items is methodologically favored. This suggests that a general demarcation line between even such distinct constructs as stress and support (or well-being) is impossible, impractical, or unnecessary.

In this regard, a comprehensive conceptual scheme should be established by tracing the roots of confusion. Some logical requirements need be articulated for a better notion of content or construct validity, which is essential for a meaningful analysis in social and behavioral sciences. Here the overlapping of constructs is